Cambridge Tracts in Mathematics and Mathematical Physics

No. 54

PACKING AND COVERING

PACKING AND COVERING

BY

C. A. ROGERS

D.Sc., F.R.S.

Astor Professor of Mathematics at University College London

CAMBRIDGE

AT THE UNIVERSITY PRESS

1964

CAMBRIDGE UNIVERSITY PRESS
Cambridge, New York, Melbourne, Madrid, Cape Town, Singapore, São Paulo, Delhi

Cambridge University Press
The Edinburgh Building, Cambridge CB2 8RU, UK

Published in the United States of America by Cambridge University Press, New York

www.cambridge.org
Information on this title: www.cambridge.org/9780521061216

First published 1964
This digitally printed version 2008

A catalogue record for this publication is available from the British Library

ISBN 978-0-521-06121-6 hardback
ISBN 978-0-521-09034-6 paperback

CONTENTS

Chapter 5. THE EXISTENCE OF REASONABLY ECONOMICAL LATTICE COVERINGS

Chapter 6. PACKINGS OF SIMPLICES CANNOT BE VERY DENSE

Chapter 7. PACKINGS OF SPHERES CANNOT BE VERY DENSE

Chapter 8. COVERINGS WITH SPHERES CANNOT BE VERY ECONOMICAL

PREFACE

Until recently the theory of packing and covering was not sufficiently well developed to justify the publication of a book devoted exclusively to it. After the publication of L. Fejes Tóth's excellent book in 1953, there would be no need for a second work on the subject, but for the fact that he confines his attention mainly to problems in two and three dimensions. Since my own interest has been mainly in the problems which arise in n-dimensional space, with n large, this book, which concentrates mainly on problems in n-dimensional space, is offered as a complement to Fejes Tóth's work. The present seems to me to be an opportune time for the publication of such a book, since I have the impression that most of the simplest general results of the subject have already been discovered, and that further progress may be rather slow, depending on detailed and complicated technical developments. If I am proved wrong in this I shall be happy.

Most mathematical authors are faced with the dilemma that, while the history of a subject usually begins with very special results and only leads gradually up to generalities, an economical and logical exposition of the subject requires that the special results should be obtained as particular cases of the general. This book is the result of a compromise between the two approaches. In the introduction we give a historical outline of the subject, stating results without proof. In the succeeding chapters we give a systematic (but by no means exhaustive) account of the general n-dimensional results of the subject.

Professor H. Davenport aroused my interest in the subjects of packing and covering when I returned to University College in 1946. I am most conscious of my debt to Professor Davenport in this as in many other ways. I must also express my gratitude to H. Minkowski, G. Voronoi, H. F. Blichfeldt and Professors H. S. M. Coxeter, H. E. Daniels, B. Delaunay, A. M. Macbeath, and Drs L. Few and G. C. Shephard, fragments of whose work will be found reproduced in scarcely disguised form in this book. Further, my own work has been much influenced by these mathematicians and by Professors E. Hlawka, K. Mahler, C. L. Siegel and Drs P. Erdös and W. Schmidt.

The idea of writing this book came to me while I was preparing my inaugural lecture on 'Packing and Covering' delivered at University College London in November 1958. The opportunity of making a first draft came when I was invited to give a course of lectures on the subject at the Canadian Mathematical Congress held in the summer of 1959 at Fredricton, New Brunswick. The final manuscript was prepared and typed while I was visiting the University of Toronto and The University of British Columbia, Vancouver for the session 1961–62 with the aid of a Canada Council grant. I am most grateful to these institutions for making this book possible.

C. A. R.

UNIVERSITY COLLEGE LONDON
January 1963

INTRODUCTION

A HISTORICAL OUTLINE OF THE THEORIES OF PACKING AND COVERING

1. Lattice packing of spheres

It is difficult to trace the first significant contribution to the mathematical theory of packing, but perhaps the honour should fall to the work of Gauss† in 1831. Although Lagrange in 1773 had developed the theory of reduction of binary quadratic forms, and found the inequalities satisfied by their coefficients, it was not until Gauss introduced the idea of a lattice in 1831 that Lagrange's results became of significance in the theory of packing.

If $\mathbf{a}_1, \mathbf{a}_2, \ldots, \mathbf{a}_n$ are n linearly independent vectors in n-dimensional Euclidean space, the set $\Lambda = \Lambda(\mathbf{a}_1, \mathbf{a}_2, \ldots, \mathbf{a}_n)$ of all vectors of the form

$$\mathbf{a} = u_1\mathbf{a}_1 + u_2\mathbf{a}_2 + \ldots + u_n\mathbf{a}_n,$$

where $u_1, u_2, \ldots, u_n$ are arbitrary integers, is called a lattice. Let $\mathbf{a}_1, \mathbf{a}_2, \ldots, \mathbf{a}_n, \mathbf{a}_{n+1}, \ldots$ be an enumeration of the points of such a lattice Λ. A system $\mathscr{K}$ consisting of the translates

$$K + \mathbf{a}_i \quad (i = 1, 2, \ldots)$$

of a given Lebesgue measurable set K, by the vectors of the lattice Λ, is called a lattice packing of K, with lattice Λ, when there is no point of space which is common to two or more of the sets of the system. Such a system $\mathscr{K}$ has a density denoted by $\rho(\mathscr{K})$, which will be defined and discussed in Chapter 1, but which may justifiably be regarded as the proportion of the whole of space covered by the sets of the packing.

When examined from this point of view, Lagrange's results imply that

$$\rho(\mathscr{K}_2) \leqslant \frac{\pi}{\sqrt{12}} = 0{\cdot}9069\ldots,$$

† These informal references, as well as more formal ones, will be found in the Bibliography. The dates quoted are, strictly, not those of the results, but those carried by the works in which they were published.

for every lattice packing $\mathscr{K}_2$ of an open circle K_2 in the Euclidean plane. It is easy to verify that

$$\rho(\mathscr{K}_2) = \frac{\pi}{\sqrt{12}},$$

in the case when K_2 is the circle

$$x_1^2 + x_2^2 < 1,$$

and $\mathscr{K}_2$ is the lattice packing of K_2 with lattice Λ generated by the points

$$(2, 0), \quad (1, \sqrt{3}).$$

Thus, if we define a lattice-packing density $\delta_L(K)$, by taking

$$\delta_L(K) = \sup_{\mathscr{K}} \rho(\mathscr{K}),$$

the supremum being over all lattice packings $\mathscr{K}$ of the set K, we have

$$\delta_L(K_2) = \frac{\pi}{\sqrt{12}} = 0{\cdot}9069\ldots, \tag{1}$$

when K_2 is the open unit circle.

Largely because of its connection with the arithmetic minimum of a positive definite quadratic form, much effort has been devoted to the study of $\delta_L(K_n)$, where K_n is the unit sphere in n-dimensional space. In 1831 Seeber, in his book on the reduction of ternary quadratic forms, established a system of inequalities satisfied by reduced forms, and in addition made a conjecture which implies that $\delta_L(K_3) = \pi/\sqrt{18}$. In his review of this book, Gauss deduced Seeber's conjecture from his inequalities, and introduced the geometric interpretation providing the connection with the lattice packings of spheres. The value of $\delta_L(K_n)$ was found by Korkine and Zolotareff in 1872 and 1877 when $n = 4$ or 5, and by Blichfeldt in 1925, 1926 and 1934 when $n = 6$, 7 and 8. In 1944 Mordell showed how the result for $n = 8$ could be very simply deduced from the case when $n = 7$. The exact value of $\delta_L(K_n)$ is unknown for $n \geqslant 9$, but lower bounds for it, which seem reasonably good and which may be exact, have been found by Chaundy in 1946 when $n = 9$ or 10, by Coxeter and Todd in 1951 when $n = 12$, and by Barnes in 1959 when $n = 11$. Some further bounds for moderately large values of n are given by Barnes and Wall (1959).

The results may be summarized in Table 1.

TABLE 1

Dimension	Density of closest lattice packing of a sphere		Reference
2	$\frac{\pi}{2\sqrt{3}}$	0·9069...	Lagrange, 1773; Gauss, 1831
3	$\frac{\pi}{3\sqrt{2}}$	0·7404...	Gauss, 1831
4	$\frac{\pi^2}{16}$	0·6168...	Korkine and Zolotareff, 1872
5	$\frac{\pi^2}{15\sqrt{2}}$	0·4652...	Korkine and Zolotareff, 1877
6	$\frac{\pi^3}{48\sqrt{3}}$	0·3729...	Blichfeldt, 1925
7	$\frac{\pi^3}{105}$	0·2952...	Blichfeldt, 1926
8	$\frac{\pi^4}{384}$	0·2536...	Blichfeldt, 1934
9	$\geqslant \frac{2\pi^4}{945\sqrt{2}}$	$\geqslant$ 0·1457...	Chaundy, 1946
10	$\geqslant \frac{\pi^5}{1920\sqrt{3}}$	$\geqslant$ 0·0920...	Chaundy, 1946
11	$\geqslant \frac{64\pi^5}{19{,}7110\sqrt{3}}$	$\geqslant$ 0·0604...	Barnes, 1959
12	$\geqslant \frac{\pi^6}{19{,}440}$	$\geqslant$ 0·0494...	Coxeter and Todd, 1951

The inequalities for $n = 9$, 10, 11 and 12 are all obtained by calculating the densities of certain carefully chosen lattice packings of spheres, the arrangement being explicitly known in each case. When n is large it seems to be necessary to fall back on indirect arguments. In 1905 Minkowski proved that for all n

$$\delta_L(K_n) \geqslant \zeta(n)/2^{n-1}, \tag{2}$$

where
$$\zeta(n) = \sum_{k=1}^{\infty} k^{-n}.$$

This is a particular case of the more general inequality

$$\delta_L(K) \geqslant \zeta(n)/2^{n-1}, \tag{3}$$

valid for any convex symmetrical body K, which was stated (by implication) by Minkowski in 1893 and proved by Hlawka in 1944. This more general inequality and some of its refinements will be discussed in §2; here we merely draw attention to the improvement

$$\delta_L(K_n) \geqslant \frac{n\zeta(n)}{e(1-e^{-n})\,2^{n-1}} \tag{4}$$

obtained by Rogers in 1947, and to its subsequent refinement by Davenport and Rogers, also in 1947.

The first significant upper bound for $\delta_L(K_n)$ was the bound

$$\frac{n+2}{2}\left(\frac{1}{\sqrt{2}}\right)^n \tag{5}$$

obtained by Blichfeldt in 1914. It was subsequently refined by Blichfeldt himself in 1929, by Rankin in 1947 and by Rogers in 1958. Although we give an account of Rogers's work in Chapter 7 it will be seen (from equation (11) of that chapter) that the improvement is slight when n is large.

In the following diagram we plot the values of the function

$$\frac{\log_e \delta_L(K_n)}{n}$$

for $n = 2, 3, \ldots, 8$, and its lower bounds for $n = 9, 10, 11, 12$, from the table, in comparison with the bounds

$$-\log_e 2 + \frac{1}{n}\log_e \frac{2n\zeta(n)}{e(1-e^{-n})}$$

and

$$-\tfrac{1}{2}\log_e 2 + \frac{1}{n}\log_e \frac{n+2}{2}$$

given by (4) and (5). Here we work with (4) and (5) rather than with any of their refinements as they are simple and explicit, while their refinements are complicated and much more difficult to calculate.

For results on 'multiple' lattice packings of spheres and circles see Few (1953), Heppes (1955, 1959) and Blundon (1963).

For results on 'mixed' packings of circles and on 'mixed' lattice packings of 3-dimensional spheres see Fejes Tóth (1953) and Few (1960).

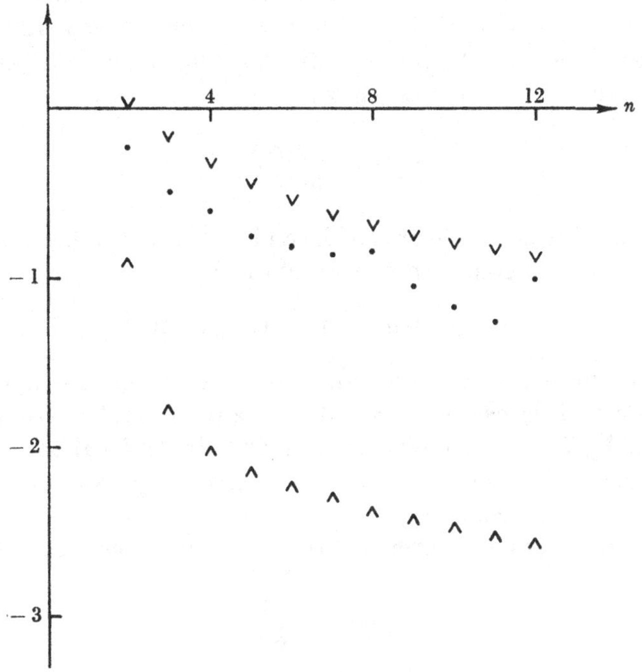

2. Lattice packing of convex sets

The first person to make a systematic study of lattice packings was Minkowski. His first interest was the theory of numbers, and his 'Fundamental Theorem', to which he was led by the study of papers of Dirichlet and Hermite on quadratic forms (see Minkowski, 1893*b*), is a useful theorem in the theory of numbers which he derived from the apparently trivial result that the density of a lattice packing is necessarily less than or equal to 1. Although he made applications of his results to the theory of numbers whenever possible, he evidently became interested in the theory of packing for its own sake. In 1904 he discussed the closest lattice packings of convex sets in three dimensions. He showed generally that in any number of dimensions the problem of determining the closest lattice packing of an asymmetrical convex set could be reduced, at least in theory, to that of determining the closest lattice packing of a symmetrical convex set. If the methods he developed for studying the case $n = 3$ are used in the case $n = 2$ it follows very simply† that the density of the closest lattice packing of a convex plane set K, with the origin $\mathbf{o}$ as centre, is

$$\delta_L(K) = \frac{4\mu(K)}{3h(K)}, \tag{6}$$

where $\mu(K)$ denotes the area of K, and $h(K)$ denotes the area of the largest hexagon, with vertices of the form

$$\mathbf{u},\ \mathbf{v},\ \mathbf{v}-\mathbf{u},\ -\mathbf{u},\ -\mathbf{v},\ -\mathbf{v}+\mathbf{u}$$

lying in this order on the boundary of K. A much more difficult discussion of the case when $n = 3$ led to a theoretical determination of $\delta_L(K)$ for a symmetrical convex 3-dimensional body K, in terms of $\mu(K)$ and the volumes of certain types of convex polyhedra inscribed in K.

In most cases the formula (6) is less convenient than the formula

$$\delta_L(K) = \frac{\mu(K)}{H(K)}, \tag{7}$$

where $H(K)$ denotes the area of the smallest hexagon or quadrilateral (necessarily symmetrical) circumscribed about the symmetrical convex plane set K. This result was discovered by Reinhardt in 1934 and rediscovered by Mahler (1947*b*).

Although Reinhardt's result gives a most satisfying answer to the problem of determining $\delta_L(K)$, when K is a convex symmetrical plane set; the problem of determining the set K of this form, for which $\delta_L(K)$ is least, remains unsolved. Both Reinhardt and Mahler (1947*b*) gave an example of such a set with

$$\delta_L(K) = \frac{8-4\sqrt{2}-\log 2}{2\sqrt{2}-1} = 0{\cdot}9024\ldots.$$

† But although the result must have been familiar to Minkowski, I have not found it in his published work.

Mahler (1946*b*) proved that all such sets satisfy

$$\delta_L(K) \geqslant \sqrt{(\tfrac{3}{4})} = 0{\cdot}8660....$$

More recently Ennola (1961) has obtained the stronger inequality

$$\delta_L(K) \geqslant \tfrac{1}{4}\{3\sqrt{2}+\sqrt{3}-\sqrt{6}\} = 0{\cdot}8813... \qquad (8)$$

and announced the lower bound 0·8925....

Despite considerable theoretical advances in the Geometry of Numbers since Minkowski's time (see Cassels, 1959), the problem of determining the value of $\delta_L(K)$ for a given convex 3-dimensional body K remains a formidable task. Minkowski himself in 1904 showed that

$$\delta_L(K) = \tfrac{18}{19} = 0{\cdot}9473...,$$

when K is the octahedron defined by

$$|x_1|+|x_2|+|x_3| < 1.$$

Whitworth (1948, 1951) used Minkowski's methods to obtain the value when K is a cube with two opposite corners truncated by plane faces and when K is a 'double cone'. Chalk in 1950 found the density when K is a 'slice' cut from a sphere. Mahler (1946*a*; see also Hlawka, 1948) obtained the density, when K is a circular cylinder, by using the theory of (general) packings of circles. This result was extended to cylinders on an arbitrary convex plane base by Chalk and Rogers (1948) and Yeh (1948), and to cylinders on a base consisting of a 3-dimensional sphere by Woods in 1958.

Apart from the results for spheres discussed in § 1, certain results for cylinders, and certain examples of space-filling sets, the exact value of $\delta_L(K)$ remains unknown for all convex sets K in 4 or more dimensions. So, when $n \geqslant 4$, the main interest lies in the determination of lower bounds for $\delta_L(K)$ for various classes of sets K and estimates for $\delta_L(K)$ for certain special sets K, the spheres and the simplices being perhaps the most interesting.

Minkowski (1893*a*) announced a result in the Geometry of Numbers concerning star bodies. Application of this result to a convex symmetrical n-dimensional set K immediately yields the inequality

$$\delta_L(K) \geqslant \zeta(n)/2^{n-1} \quad \left(\zeta(n) = \sum_{k=1}^{\infty} k^{-n}\right). \tag{9}$$

Minkowski only published a proof of his result in 1905 in the special case when K is a sphere or ellipsoid. Blichfeldt† in 1919 stated that he had obtained a stronger inequality in the case when K is convex and symmetrical. In 1944 Hlawka published the first complete proof of Minkowski's theorem, and Mahler published (a few months later) a proof of a result which was only slightly weaker. In 1945 Siegel published a proof making use of the rather deep considerations that Minkowski had used in discussing the case of an ellipsoid; it seems likely that Siegel's proof is similar to Minkowski's unpublished work; it has the advantage that it works with a natural measure in the space of lattices. Since 1945 many proofs and refinements‡ of the Minkowski–Hlawka theorem have been published. The first refinement was obtained by Mahler; most of the more recent ones are due to Rogers and Schmidt. Schmidt (1958) was the first to obtain an inequality of the form

$$\delta_L(K) \geqslant cn/2^n, \tag{10}$$

valid for all n and a suitable constant c; in his latest paper on the subject he shows that this inequality holds for convex symmetrical K when n is sufficiently large provided

$$c < \log 2. \tag{11}$$

The proofs of the more refined inequalities are exceedingly complicated. In Chapter 4 we give only the simplest approach to the problem, proving that

$$\delta_L(K) \geqslant 1/2^{n-1} \tag{12}$$

for each convex symmetrical n-dimensional set K. Since

$$\zeta(n) = 1 + O((\tfrac{1}{2})^n)$$

† See also Bernstein (1918).

‡ Bateman (1962), Cassels (1953), Davenport and Rogers (1947), Lekkerkerker (1956), Macbeath and Rogers (1955, 1958 *a*, *b*), Mahler (1946), Malyšev (1952), Rogers (1947, 1951 *a*, 1954, 1955 *a*, *b*, 1956 *a*, *b*, 1957 *b*, 1958 *b*), Sanov (1952), Santaló (1950), Schmidt (1956 *a*, *b*, 1957, 1958 *a*, *b*, 1959, 1963), Weil (1946).

this inequality is only fractionally weaker than (9) when n is large.

There remains a wide gap between the results of the Minkowski–Hlawka type, showing that for each convex symmetrical set K the lattice-packing density is at least $cn/2^n$, and the results of Blichfeldt type, showing that the lattice-packing density (and indeed the packing density) is less than $cn/(\sqrt{2})^n$ for an n-dimensional sphere.

So far we have been mainly concerned with the lattice packing of symmetrical convex sets. When we turn to convex sets that are not necessarily symmetrical, we find that despite Minkowski's reduction of the general case to the symmetrical case very little is known. If K is any set, its difference set $\mathsf{D}K$ is defined to be the set of all points of the form

$$\mathbf{x}-\mathbf{y}$$

with $\mathbf{x}$ and $\mathbf{y}$ in K. It is easy to verify that $\mathsf{D}K$ is automatically symmetrical in $\mathbf{o}$ and that $\mathsf{D}K$ is convex if K is convex. Provided K is convex, Minkowski's argument (see Chapter 6, Theorem 6.7) shows that

$$\frac{\delta_L(K)}{\mu(K)} = \frac{\delta_L(\mathsf{D}K)}{\mu(\mathsf{D}K)}2^n. \tag{13}$$

In 1904 Minkowski claimed to have determined $\delta_L(K)$, when K is a tetrahedron, in this way, but he was plainly wrong in asserting that the difference body of a tetrahedron was an octahedron.†

When $n = 2$ the relationship between K and $\mathsf{D}K$ is not difficult to investigate, and Fáry in 1950 was able to show that, for all open convex plane sets K,

$$\delta_L(K) \geqslant \tfrac{2}{3},$$

with equality only when K is a triangle. For $n \geqslant 3$ the key to progress is to find the upper bound of the ratio

$$\frac{\mu(\mathsf{D}K)}{\mu(K)}.$$

† For Pepper's discussion of this point see Hancock (1939).

In 1925 Rademacher showed that the bound was 6 when n is 2. In 1928 Estermann and Süss independently obtained the result 20 when $n = 3$. The natural conjecture is that the exact upper bound is the binomial coefficient

$$\binom{2n}{n} = \frac{(2n)!}{(n!)^2},$$

in general, and this was proved by Rogers and Shephard in 1957 and in a more geometrical way in 1958 (see Chapter 2, §2). Combining this with (12) and (13) we immediately have

$$\delta_L(K) \geqslant \frac{2(n!)^2}{(2n)!} \tag{14}$$

for any convex set K; we obtain this inequality in a slightly different way in Theorem 4.4 of Chapter 4.

Since

$$\frac{2(n!)^2}{(2n)!} \sim \frac{2\sqrt{(\pi n)}}{4^n} \quad (n \to \infty),$$

the inequality (14) is very weak when n is large. Rogers and Shephard (1957) remark that the right-hand side of any equality of the form (14) is necessarily very small, since in the case of a simplex

$$\delta_L(K) \leqslant \frac{2^n(n!)^2}{(2n)!} \sim \frac{\sqrt{(\pi n)}}{2^n} \quad (n \to \infty). \tag{15}$$

An account of this is given in Chapter 6.

3. Packing of convex sets

As a natural extension of the ideas of §1 a system $\mathscr{K}$ consisting of the translates

$$K + \mathbf{a}_i \quad (i = 1, 2, \ldots)$$

of a given Lebesgue measurable set K, by the vectors of a sequence $\mathbf{a}_1$, $\mathbf{a}_2$, ..., which may be finite or infinite, is called a packing of K when there is no point of space which is common to two or more of the sets of the system. If the vectors $\mathbf{a}_1$, $\mathbf{a}_2$ ... are distributed in a sufficiently regular way throughout the whole of space (in particular, for example, if the set of points $\{\mathbf{a}_i\}$ is periodic with some period in each coordinate) we can assign a density $\rho(\mathscr{K})$ to the system. For a detailed discussion see

Chapter 1; here it suffices to remark that this density is consistent with our previous density, when $\mathbf{a}_1, \mathbf{a}_2, \ldots$ is an enumeration of the points of a lattice, and that as before the density of the system is a mathematical construct, which corresponds to the intuitive notion of the proportion of the whole of space covered by the sets of the packing. We define the packing constant $\delta(K)$ of a Lebesgue measurable set K, by taking

$$\delta(K) = \sup_{\mathscr{K}} \rho(\mathscr{K}),$$

the supremum being over all the packings $\mathscr{K}$ for which the density $\rho(\mathscr{K})$ is defined.†

The first mathematician to study packings, which are not necessarily lattice packings, seems to have been Thue. In a lecture at the Scandinavian Natural Science Congress in 1892, he gave an account of a result showing that the packing density $\delta(K_2)$ of a circle is equal to the density

$$\frac{\pi}{\sqrt{12}} = 0{\cdot}9069\ldots$$

attained for the hexagonal arrangement, where each circle touches exactly six others. The published account of the lecture is very short and is hardly sufficient to enable his proof to be reconstructed.‡

Rather later, in 1910, Thue gave another proof using very different ideas. Although this proof is convincing, it does seem to be open to the objection that it is no easy task to establish certain compactness results which he takes for granted.§

In 1940 Fejes Tóth¶ gave a new proof of Thue's result, and Segre and Mahler gave another independently a few months later in the year. These two papers also obtained results concerning finite packings of circles into polygons with at most six

† This is not the same as, but is simpler than and equivalent to the formal definition of $\delta(K)$ given in Chapter 1.

‡ In particular, there is no indication of the considerations he presumably must have used to discuss the situation when seven of his points lie within a distance $2k/\sqrt{3}$ of one of the points.

§ Compare Thue's 1910 paper with the work of Oler (1961).

¶ We use Fejes Tóth's full name throughout, although his earlier papers appeared under the name L. Fejes.

sides. We give a proof of Thue's result in Chapter 7; see the case $n = 2$ of Theorem 7.1.

In 1950 Fejes Tóth showed that

$$\delta(K) \leqslant \frac{\mu(K)}{H(K)}, \tag{16}$$

where $H(K)$ denotes the area of the smallest hexagon circumscribed to K, for each convex two-dimensional set K. Since it is easy to verify that

$$\delta(K) \geqslant \frac{\mu(K)}{H(K)},$$

when the minimal circumscribing hexagon is symmetrical, which it certainly is when K is symmetrical, the result is best possible in many cases. In fact, Fejes Tóth's work went considerably beyond (16), since he was able to obtain the same inequality for the density of non-overlapping arrangements of general affine transforms of K (not merely of translates of K), subject to the condition that all the transforms should have the same area. We warn the reader that one (at least) of Fejes Tóth's steps is not exactly easy; he should consult Fejes Tóth (1953*a*), pp. 86–7, Cassels (1959), pp. 240–3, and Davies† (——).

About the same time‡ Rogers (1951*b*) proved that

$$\delta(K) = \delta_L(K),$$

for any plane convex set K. Combining this with Reinhardt's result (7) of §2 we have

$$\delta(K) = \delta_L(K) = \frac{\mu(K)}{H(K)} \tag{17}$$

in the special case when K is symmetrical.

We give no account of this work, as it is largely covered by the excellent monograph of Fejes Tóth (1953*a*); but see also the refinements of Oler (1961), Molnár (1961) and Davies (——), and the survey of Zassenhaus (1961).

Little precise is known about the packing density $\delta(K)$ in

† References of this form are to work unpublished at the time the proofs were corrected; the available details are given in the bibliography.

‡ The result was announced, a little before it was proved to the complete satisfaction of the author, in Chalk and Rogers (1948); see also the corrigendum Rogers (1960), which does not however affect the result quoted.

three or more dimensions. Naturally $\delta(K) = 1$, if translates of K can be fitted together without overlapping to fill the whole of space, apart from sets of measure zero, and something is known of the convex sets K with this property. They have been completely enumerated when $n = 3$ and the translations are restricted to those of a lattice (see, for example, Coxeter, 1962*a*). Also it is easy to show that

$$\delta(C) = \delta(K)$$

if C is a cylinder with the convex two-dimensional base K.

Although Blichfeldt in his 1914 paper did consider 'generalized lattices', it does not seem to be possible to deduce a good estimate for $\delta(K_n)$, where K_n is an n-dimensional sphere, from this paper. However, when he returned to the subject in 1929, he adopted a slightly different point of view, and explicitly considered general packings of spheres, proving in particular that

$$\delta(K_n) \leqslant \frac{n+2}{2}\left(\frac{1}{\sqrt{2}}\right)^n. \tag{18}$$

He also gave a refinement of this result, which was subsequently improved by Rankin (1947) and Rogers (1958*c*). We give a proof of Rogers's inequality

$$\delta(K_n) \leqslant \sigma_n \tag{19}$$

in Chapter 7. Since

$$\sigma_2 = \pi/\sqrt{12}$$

and

$$\sigma_n < \frac{n+2}{2}\left(\frac{1}{\sqrt{2}}\right)^n,$$

this includes the results of Thue and Blichfeldt.

When $n = 3$ the inequality (19) becomes

$$\delta(K_3) \leqslant (\sqrt{18})(\cos^{-1}\tfrac{1}{3} - \tfrac{1}{3}\pi) = 0\cdot7797\ldots,$$

which seems to be the strongest inequality so far published, despite a widespread belief that

$$\delta(K_3) = \delta_L(K_3) = \frac{\pi}{\sqrt{18}} = 0\cdot7404\ldots.$$

For some interesting work on this problem see Fejes Tóth (1952) and Coxeter (1958).

Blichfeldt (1936), Hlawka (1945, 1947), and Rankin (1949*a*, *b*,

1948) also obtained results for the packings of the set defined by the inequality

$$|x_1|^p + |x_2|^p + \ldots + |x_n|^p < 1,$$

where $p \geqslant 1$ is a fixed parameter.

Since trivially

$$\delta_L(K) \leqslant \delta(K),$$

the lower bounds for $\delta_L(K)$, discussed in §2, are *a fortiori* lower bounds for $\delta(K)$. One might imagine that it was easier to obtain lower bounds for $\delta(K)$ than for $\delta_L(K)$. But this seems to be true only to a limited extent. It is easy to obtain the lower bounds

$$\frac{2(n!)^2}{(2n)!} \quad \text{and} \quad \frac{1}{2^{n-1}}$$

for $\delta(K)$, in the cases when K is convex and when K is convex and symmetrical. Indeed, in the special case of an n-dimensional sphere K_n, we are able to show that

$$\delta(K_n) \geqslant cn/2^n,$$

where c is a constant, but this result is rather weaker than the results mentioned in §2. We give details in Chapter 2. However, the more refined methods of obtaining dense packings of convex sets all rely on the theory of lattice packings and depend on the construction of dense lattice packings. Consequently these methods all provide the same lower bounds for $\delta_L(K)$ as for $\delta(K)$.

The author knows of no example for which

$$\delta_L(K) < \delta(K).$$

Although he is sure that

$$\delta_L(K_3) = \delta(K_3),$$

this has never been proved. He believes that probably

$$\delta_L(K_4) = \delta(K_4),$$

but thinks it quite possible that

$$\delta_L(K_5) < \delta(K_5)$$

or that

$$\delta_L(K_7) < \delta(K_7).$$

He conjectures that $\delta_L(K_n) < \delta(K_n)$

for all sufficiently large n.

One general result is worth noticing. The proof of Minkowski that

$$\frac{\delta_L(K)}{\mu(K)} = \frac{\delta_L(\mathsf{D}K)}{\mu(\mathsf{D}K)} 2^n, \tag{20}$$

when K is convex (see (13) of §2), extends without difficulty to show that

$$\frac{\delta(K)}{\mu(K)} = \frac{\delta(\mathsf{D}K)}{\mu(\mathsf{D}K)} 2^n, \tag{21}$$

when K is convex, thus theoretically reducing the problem for convex asymmetrical sets to that for convex symmetrical ones.

The work of Rogers and Shephard, referred to in §2, applies equally to general packings of simplices, showing that

$$\frac{2(n!)^2}{(2n)!} \leqslant \delta_L(K) \leqslant \delta(K) \leqslant \frac{2^n(n!)^2}{(2n)!},$$

when K is an n-dimensional simplex. An account of this is given in Chapter 6.

For definitions and general results on the theory of packings, the reader is referred to the work of Rado (1949, 1951), Hlawka (1949), and (if he can possibly obtain a copy) Davenport (1955).

Works on the Geometry of Numbers naturally contain much that is related to and relevant to the theory of packing; the reader is particularly referred to the recent books of Keller (1954) and Cassels (1959). For an excellent account of much related work, mainly in two and three dimensions, see the book of Fejes Tóth (1953a) and those papers of his listed in the bibliography to this present book; see also the work of Heppes and Molnár. There are a few other papers rather closely related to the subject of this book which require special mention here. Rankin (1955) studies the packing spherical caps on the surface of a sphere. Few (1953, ——) studies multiple packings of spheres in n-dimensions. Smith (1951) obtained a striking result on arrangements of points in the plane with the property that no two points x, y have the property that

$$|(y_1 - x_1)(y_2 - x_2)| \leqslant 1.$$

4. Covering

The mathematical theory of the type of coverings, to which we restrict ourselves, was developed rather later than the corresponding theory for packings, despite the similarities between the two concepts. We say that a system $\mathscr{K}$ of translates

$$K+\mathbf{a}_i \quad (i = 1, 2, \ldots)$$

of a set K by a sequence of points $\mathbf{a}_1$, $\mathbf{a}_2$, ... forms a covering, if each point of space lies in at least one of the sets of the system. If the points $\mathbf{a}_1$, $\mathbf{a}_2$, ... are an enumeration of the points of a lattice, the covering is called a lattice covering. Provided K is bounded and measurable in the Lebesgue sense, and the sequence of points is sufficiently regularly distributed throughout the space, we can associate a density with the system $\mathscr{K}$. This density is defined in Chapter 1 and will be denoted by $\rho(\mathscr{K})$. It may be described as the limiting ratio of the sum of the measures† of those sets of the system $\mathscr{K}$, which lie in a large cube, to the measure of the cube, as it becomes infinitely large.

If K is bounded and Lebesgue measurable its covering density $\vartheta(K)$, and its lattice covering density $\vartheta_L(K)$ may be defined as the lower bounds of $\rho(\mathscr{K})$, taken respectively over all coverings $\mathscr{K}$ and over all lattice coverings $\mathscr{K}$.

Although earlier results may lie concealed, or merely ignored, in the vast literature of mathematics, the earliest result in the subject seems to be the proof of Kershner in 1939 that

$$\vartheta(K_2) = \vartheta_L(K_2) = \frac{2\pi}{3\sqrt{3}} = 1{\cdot}209..., \tag{22}$$

where K_2 is a (closed) circle. Here we have

$$\rho(\mathscr{K}_2) = \vartheta(K_2)$$

in the case when $\mathscr{K}_2$ is a system of circles, with their centres at the points of a hexagonal lattice, and their common radius chosen so that they just cover the plane. A proof of Kershner's result is given in Chapter 8, in the case $n = 2$ of Theorem 8.

† Note it is important to work with the sum of the measures of the sets rather than the measure of the union of the sets; otherwise the density would be unity in all normal cases.

In 1946 Fejes Tóth used a result of Sas (1939), concerning the symmetrical hexagons inscribed in a symmetrical plane convex set, to show that

$$\vartheta_L(K) \leqslant \frac{2\pi}{3\sqrt{3}},$$

so that the circle is the least economical symmetrical convex plane set, from the point of view of lattice coverings.

In 1950 Fáry studied lattice coverings, with a plane convex set K, which is not necessarily symmetrical. His results immediately imply that

$$\vartheta_L(K) = \frac{\mu(K)}{h_s(K)}$$

for each convex plane set K, where $h_s(K)$ denotes the area of the largest centrally symmetric hexagon contained in K. He proved also that, for each convex plane set K,

$$\vartheta_L(K) \leqslant 1{\cdot}5,$$

with equality only when K is a triangle.

In the same year Fejes Tóth† succeeded in proving that, for all convex plane sets K,

$$\vartheta_L(K) \geqslant \vartheta(K) \geqslant \frac{\mu(K)}{h(K)},$$

where $h(K)$ denotes the area of the largest hexagon inscribed in K. Since $h_s(K) = h(K)$, when K is symmetrical, he deduced that

$$\vartheta_L(K) = \vartheta(K) = \frac{\mu(K)}{h_s(K)} \tag{23}$$

in this case. The question of whether

$$\vartheta_L(K) = \vartheta(K)$$

for all convex plane sets K remains open.

When we turn to problems of covering in three dimensions we find that very little is known. Bambah (1954a) discussed lattice coverings with 3-dimensional spheres and showed that

$$\vartheta_L(K_3) = \frac{5\sqrt{5}}{24}\pi = 1{\cdot}464\ldots. \tag{24}$$

† See also Bambah and Rogers (1952) for some additional comments and explanations.

Simpler proofs were given by Barnes (1956) and Few (1956). More recently, in 1959, Coxeter, Few and Rogers obtained a general result (see Chapter 8) which implies that

$$1{\cdot}431\ldots = \frac{3\sqrt{3}}{2}(3\sec^{-1}3-\pi) \leqslant \vartheta(K_3) \leqslant \vartheta_L(K_3) = 1{\cdot}464\ldots, \tag{25}$$

so that $\vartheta(K_3)$ is certainly close to the conjectured value 1·464....

Bambah (1954*b*) gives an example to show that

$$\vartheta_L(K_4) \leqslant \frac{2}{5\sqrt{5}}\pi^2 = 1{\cdot}765\ldots. \tag{26}$$

It follows, from this and the work of Coxeter, Few and Rogers, that

$$1{\cdot}658\ldots = \frac{192}{5\sqrt{5}}\pi(\tfrac{1}{2}\sec^{-1}4-\tfrac{1}{5}\pi) \leqslant \vartheta(K_4) \leqslant \vartheta_L(K_4) \leqslant 1{\cdot}765\ldots. \tag{27}$$

These inequalities appear to leave some chance for the existence of a lattice covering with 4-dimensional spheres, with density less than the density

$$\frac{3}{5\sqrt{5}}\pi^2 = 1{\cdot}765\ldots,$$

which Bambah conjectured was the least possible.

Early attempts to construct or establish the existence of economical coverings of space with convex bodies, by Rogers (1950), Bambah and Roth (1952), and Rogers (1958*a*, *b*), and with spheres by Davenport (1951, 1952) and Watson (1956), were relatively unsuccessful. These authors all obtained results of the form

$$\vartheta_L(K) \leqslant c^n$$

for all sufficiently large n, for suitable values of $c > 1$, and for suitable restrictions on the convex n-dimensional set K. The smallest value of c obtained was the value 1·017... of Watson. However, Rogers (1957*a*) found a rather different approach to the subject; he showed that

$$\vartheta(K) \leqslant n\log n + n\log\log n + 5n \tag{28}$$

for $n \geqslant 3$, for all convex n-dimensional sets K. But it was not until 1959 that he succeeded in modifying his method, in a way which enabled him to prove the existence of economical lattice coverings of space. His closest estimates,

$$\vartheta_L(K) \leqslant n^{\log_2 \log_e n + c}, \tag{29}$$

for some constant c and all convex n-dimensional sets K, and

$$\vartheta_L(K_n) \leqslant cn\,(\log_e n)^{\frac{1}{2}\log_2 2\pi e}, \tag{30}$$

for some constant c and an n-dimensional sphere K_n, are rather difficult to establish, depending on the work of Schmidt (1958*b* or 1959). In Chapter 3 we give an account of the rather simple proof of (28). In Chapter 5 we give proofs of the inequalities

$$\vartheta_L(K) \leqslant n^{\log_2 n + c \log\log n}, \tag{31}$$

$$\vartheta_L(K_n) \leqslant c(n \log_2 n)^{\frac{1}{2}\log_2 2\pi e} \tag{32}$$

which are rather weaker than (29) and (30), but which are much easier to prove. The theorems, unlike those of Davenport and Watson, depend on existence proofs and do not yield any explicit constructions for economical lattice coverings with spheres.

The first lower bounds for the covering densities of an n-dimensional sphere were the results

$$\vartheta_L(K_n) \geqslant \tfrac{4}{3} - \epsilon_n,$$

where $\epsilon_n \to 0$ as $n \to \infty$, of Bambah and Davenport (1952) and

$$\vartheta(K_n) \geqslant \tfrac{16}{15} - \epsilon_n,$$

where $\epsilon_n \to 0$ as $n \to \infty$, of Erdös and Rogers (1953). Quite recently Coxeter, Few and Rogers, in work mentioned above, and described in Chapter 8, have shown that

$$\vartheta_L(K_n) \geqslant \vartheta(K_n) \geqslant \tau_n, \tag{33}$$

where $$\tau_n \sim \frac{n}{e\sqrt{e}} \quad (n \to \infty).$$

For some general results on the theory of coverings the reader should see Hlawka (1949), Bambah (1953) and Davenport

(1955). Fejes Tóth's book (1953a) contains a number of results in and related to the theory in two and three dimensions; see also his more recent papers. A paper by Blundon (1957) studies multiple lattice coverings of the plane by circles. In a recent paper Erdös and Rogers (1962) establish the existence of economical coverings of space with convex bodies having the additional property that no point of space is covered too often. Bambah, Rogers and Zassenhaus hope to publish an alternative proof of Fejes Tóth's result (23) shortly.

Chapter 1

PACKING AND COVERING DENSITIES

1. Introduction

The concept of a covering is one of the simplest and most general in Mathematics. In the terminology and symbolism of abstract set theory, the system of sets $S_1, S_2, \ldots$ is said to cover the set S, if

$$\bigcup_i S_i \supset S,$$

i.e. if each element of S belongs to at least one of the sets $S_1, S_2, \ldots$. It is not surprising that such a concept is much used in various mathematical theories. The concept of a packing is hardly less simple nor less ubiquitous. The system of sets $S_1, S_2, \ldots$ is said to form a packing into the set S, if

$$S_i \cap S_j = \varnothing \quad (i \neq j), \quad \bigcup_i S_i \subset S,$$

i.e. if no two of the sets $S_1, S_2, \ldots$ have any element in common, and each element of each of the sets $S_1, S_2, \ldots$ belongs to S.

However, in this monograph we shall only be concerned with very special packings and coverings. The set S will be a set of points in n-dimensional Euclidean space, and will be usually either the whole of this space or a cube with its edges parallel to the coordinate axes; the sets $S_1, S_2, \ldots$ will be a finite or countably infinite system of 'translates' of a single set K, which will usually be convex. We use the usual vector notation and say that the set of all points of the form

$$\mathbf{k}+\mathbf{a},$$

where $\mathbf{k} \in K$, and $\mathbf{a}$ is a fixed point or vector, is the translate of K by the vector $\mathbf{a}$; we use $K+\mathbf{a}$ to denote this translate.

Instead of repeatedly talking about a packing of translates of K into the whole of space we shall simply say a packing of K, and similarly a covering with K will mean a covering of the whole space by translates of K.

If $\{\mathbf{a}_i\}$ is a sequence of points, and K is a set with a finite positive Lebesgue measure $\mu(K)$, we can introduce the concepts of the upper and lower densities of the system $\mathscr{K}=\{K+\mathbf{a}_i\}$ of sets $K+\mathbf{a}_1$, $K+\mathbf{a}_2$, ... We use C to denote a half-open half-closed 'cube' with its edges parallel to the coordinate axes, i.e. a set defined by inequalities of the form

$$\left.\begin{array}{l} c_1-\tfrac{1}{2}s \leqslant x_1 < c_1+\tfrac{1}{2}s, \\ c_2-\tfrac{1}{2}s \leqslant x_2 < c_2+\tfrac{1}{2}s, \\ \cdots\cdots\cdots\cdots\cdots\cdots\cdots\cdots \\ c_n-\tfrac{1}{2}s \leqslant x_n < c_n+\tfrac{1}{2}s, \end{array}\right\} \qquad (1)$$

where $c_1, c_2, \ldots, c_n$ are the coordinates of a point, the centre of the cube, and s is a positive number, the edge-length of the cube. We write

$$\rho_+(\mathscr{K}, C) = \frac{1}{\mu(C)} \sum_{(K+\mathbf{a}_i)\cap C\neq\emptyset} \mu(K+\mathbf{a}_i),$$

$$\rho_-(\mathscr{K}, C) = \frac{1}{\mu(C)} \sum_{(K+\mathbf{a}_i)\subset C} \mu(K+\mathbf{a}_i).$$

Thus $\rho_+(\mathscr{K}, C)$ is the ratio of the total measure of those sets of the system $\mathscr{K}$ which have a point in common with the cube C, to the measure of C; while $\rho_-(K, C)$ is the ratio of the total measure of those sets of the system $\mathscr{K}$ which are contained in the cube C, to the measure of C. Using $s(C)$ to denote the length of the edge (or side) of C, we define the upper and lower densities of the system $\mathscr{K}$ by the formulae

$$\begin{aligned} \rho_+(\mathscr{K}) &= \limsup_{s(C)\to\infty} \rho_+(\mathscr{K}, C) \\ &= \lim_{s\to\infty} \sup_{s(C)\geqslant s} \rho_+(\mathscr{K}, C), \\ \rho_-(\mathscr{K}) &= \liminf_{s(C)\to\infty} \rho_-(\mathscr{K}, C) \\ &= \lim_{s\to\infty} \inf_{s(C)\geqslant s} \rho_-(\mathscr{K}, C). \end{aligned}$$

It follows immediately from these definitions that, provided K has finite positive measure,

$$\rho_-(\mathscr{K}) \leqslant \rho_+(\mathscr{K}). \qquad (2)$$

The following theorems are scarcely less immediate.

THEOREM 1.1. *If K is bounded with a positive measure, and the system $\mathscr{K}$ of translates of K forms a packing,*

$$\rho_+(\mathscr{K}) \leqslant 1. \tag{3}$$

THEOREM 1.2. *If K is bounded with a positive measure, and the system $\mathscr{K}$ of translates of K forms a covering,*

$$\rho_-(\mathscr{K}) \geqslant 1. \tag{4}$$

Proof of Theorem 1.1. Let $s(K)$ denote the edge-length of any fixed cube, with edges parallel to the coordinate axes, which contains K. Indeed we shall adopt this conventional notation throughout the book: if K is a cube of the form (1), then $s(K)$ will denote its edge-length; but in general $s(K)$ will merely denote the edge-length of some such cube containing K.

If C is a cube, the sets

$$K + \mathbf{a}_i \tag{5}$$

that have a point in common with C all lie in the cube C', concentric with C and with edge-length $s(C)+2s(K)$. Since all the sets (5) are disjoint, it follows that

$$\sum_{(K+\mathbf{a}_i)\cap C \neq \emptyset} \mu(K+\mathbf{a}_i) \leqslant [s(C)+2s(K)]^n,$$

so that

$$\rho_+(\mathscr{K}, C) \leqslant \left[1+\frac{2s(K)}{s(C)}\right]^n.$$

Hence

$$\rho_+(\mathscr{K}) \leqslant \limsup_{s(C)\to\infty} \left[1+\frac{2s(K)}{s(C)}\right]^n = 1.$$

Proof of Theorem 1.2. Let $s(K)$ denote the edge-length of any fixed cube, with edges parallel to the coordinate axes, which contains K. Then if C is any cube with $s(C) > 2s(K)$, each point of the cube C'' concentric with C with side $s(C)-2s(K)$ lies in one of the sets $K+\mathbf{a}_i$, and this set $K+\mathbf{a}_i$ is necessarily contained in C. Thus

$$C'' \subset \bigcup_{K+\mathbf{a}_i \subset C} \{K+\mathbf{a}_i\}$$

and so

$$\{s(C)-2s(K)\}^n \leqslant \sum_{K+\mathbf{a}_i \subset C} \mu(K+\mathbf{a}_i).$$

Hence, provided $s(C) > 2s(K)$, we have

$$\rho_-(\mathscr{K}, C) \geqslant \left\{1 - \frac{2s(K)}{s(C)}\right\}^n.$$

Consequently,

$$\rho_-(\mathscr{K}) \geqslant \liminf_{s(C)\to\infty} \left\{1 - \frac{2s(K)}{s(C)}\right\}^n = 1.$$

Now, with each bounded set K with a positive measure, we associate two densities. The *packing density* $\delta(K)$ of K is defined by the formula

$$\delta(K) = \sup_{\mathscr{K}} \rho_+(\mathscr{K}), \tag{6}$$

the supremum being taken over all systems $\mathscr{K}$ of translates of K, which form packings into the whole of space. The covering density $\vartheta(K)$ of K is defined similarly by the formula

$$\vartheta(K) = \inf_{\mathscr{K}} \rho_-(\mathscr{K}), \tag{7}$$

the infimum being taken over all systems $\mathscr{K}$ of translates of K, which form a covering of the whole space. By these definitions and Theorems 1.1 and 1.2, we have the following result.

THEOREM 1.3. *If K is bounded with a positive measure, its packing and covering densities satisfy*

$$\delta(K) \leqslant 1 \leqslant \vartheta(K). \tag{8}$$

It is not difficult to see that $\delta(K)$ and $\vartheta(K)$ are affine invariants of K; this is shown in the next two sections of this chapter.

But we are also interested in a more restricted type of packing and covering, the lattice packings and the lattice coverings. If $\mathbf{a}_1, \mathbf{a}_2, \ldots, \mathbf{a}_n$ are n linearly independent vectors in n-dimensional space, the set of all points of the form

$$\mathbf{a} = u_1\mathbf{a}_1 + u_2\mathbf{a}_2 + \ldots + u_n\mathbf{a}_n,$$

where $u_1, u_2, \ldots, u_n$ are arbitrary integers, will be called a lattice. A system $\mathscr{K}$ of translates $\{K + \mathbf{a}_i\}$ of a given set K is called a lattice system, if the points $\mathbf{a}_1, \mathbf{a}_2, \ldots$, are an enumeration of the points of a lattice Λ. The lattice Λ is called the lattice of the system. A lattice system $\mathscr{K}$ is called a lattice packing of K if it is a packing into the whole of space; it is called a lattice covering with K if it is a covering of the whole of space. If K is a bounded set with positive measure, the lattice-packing density $\delta_L(K)$ and the lattice-covering density $\vartheta_L(K)$ are defined by the formulae

$$\delta_L(K) = \sup_{\mathscr{K}_L} \rho_+(\mathscr{K}_L), \tag{9}$$

$$\vartheta_L(K) = \inf_{\mathscr{K}_L} \rho_-(\mathscr{K}_L), \tag{10}$$

the supremum being taken over all lattice packings $\mathscr{K}_L$ of K and the infimum being taken over all lattice coverings $\mathscr{K}_L$ with K.

As a technical convenience we also introduce the idea of periodic packings and coverings. Let $\mathbf{a}_1, \mathbf{a}_2, \ldots, \mathbf{a}_N$ be a set of points and let $\mathbf{b}_1, \mathbf{b}_2, \ldots$ be the points of the lattice of all points that have integral multiples of a positive constant s for coordinates. Then the system $\mathscr{K}_P$ of translates

$$K + \mathbf{a}_i + \mathbf{b}_j \quad (i = 1, 2, \ldots, N;\ j = 1, 2, \ldots)$$

of a set K is periodic in each of the coordinates with period s. Such a system of translates will be called a periodic system. A periodic system, which is a packing or covering, will be called a periodic packing or a periodic covering. If K is a bounded set with positive measure, the periodic packing density $\delta_P(K)$ and the periodic covering density $\vartheta_P(K)$ are defined by the formulae

$$\delta_P(K) = \sup_{\mathscr{K}_P} \rho_+(\mathscr{K}_P), \tag{11}$$

$$\vartheta_P(K) = \inf_{\mathscr{K}_P} \rho_-(\mathscr{K}_P), \tag{12}$$

the supremum being taken over all periodic packings $\mathscr{K}_P$ of K, and the infimum being taken over all periodic coverings $\mathscr{K}_P$. We shall see later in this chapter that

$$\delta_P(K) = \delta(K), \quad \vartheta_P(K) = \vartheta(K).$$

Comparing the various definitions, and using Theorem 1.3, we obtain

THEOREM 1.4. *If K is bounded with a positive measure its various packing and covering densities satisfy*

$$\delta_L(K) \leqslant \delta_P(K) \leqslant \delta(K) \leqslant 1 \leqslant \vartheta(K) \leqslant \vartheta_P(K) \leqslant \vartheta_L(K). \tag{13}$$

In the next section we show that $\delta_L(K)$ and $\delta(K)$ are affine invariants of K and prove that $\delta_P(K) = \delta(K)$. In the following section we state the corresponding results for coverings, which can be proved by using the same methods.

2. Invariance of the packing density

In this section our main aim is to establish the invariance of the packing density of a set under an affine transformation. It is, however, convenient to give first some lemmas which enable us to calculate the upper and lower densities of certain rather special types of systems of translates.

THEOREM 1.5. *Let K be a bounded set with positive measure, let C be a cube (with its edges parallel to the coordinate axes) with edge-length $s(C)$, and let T be a non-singular affine transformation. Let $\mathbf{a}_1, \mathbf{a}_2, \ldots, \mathbf{a}_N$ be a set of points, and let $\mathbf{b}_1, \mathbf{b}_2, \ldots$ be the points of the lattice of all points that have integral multiples of $s(C)$ for co-ordinates. Let $\mathscr{K}_P$ be the periodic system of sets*

$$K + \mathbf{a}_i + \mathbf{b}_j \quad (i = 1, 2, \ldots, N; j = 1, 2, \ldots), \tag{14}$$

and let $\mathsf{T}\,\mathscr{K}_P$ denote the system of sets

$$\mathsf{T}(K + \mathbf{a}_i + \mathbf{b}_j) \quad (i = 1, 2, \ldots, N; j = 1, 2, \ldots). \tag{15}$$

Then

$$\rho_+(\mathsf{T}\mathscr{K}_P) = \rho_-(\mathsf{T}\mathscr{K}_P) = \rho_+(\mathscr{K}_P) = \rho_-(\mathscr{K}_P) = \frac{N\mu(K)}{\mu(C)}. \tag{16}$$

Proof. We first note that the addition of any vector of the lattice of points that have integral multiples of $s(C)$ for co-ordinates to one of the points $\mathbf{a}_1, \mathbf{a}_2, \ldots, \mathbf{a}_N$ leaves the systems $\mathscr{K}_P$, $\mathsf{T}\mathscr{K}_P$ unchanged, apart from the orders in which their translates are enumerated. So we may suppose that vectors from this lattice have been added to the points $\mathbf{a}_1, \mathbf{a}_2, \ldots, \mathbf{a}_N$ in a way that ensures that each set $K + \mathbf{a}_i$ has at least one point lying in the cube C.

We now remark that, since T is a non-homogeneous linear transformation, the system $\mathsf{T}\mathscr{K}_P$ is the system of translates of the set $\mathsf{T}K$ by the vectors

$$\mathsf{T}(\mathbf{a}_i + \mathbf{b}_j) - \mathsf{T}\mathbf{o} \quad (i = 1, 2, \ldots, N; j = 1, 2, \ldots).$$

Let G denote a cube of the type (1) with a large edge-length satisfying

$$s(G) > 2s(\mathsf{T}C)+2s(\mathsf{T}K).$$

Let G' and G'' denote the cubes concentric with G having edge-lengths

$$s(G)-2s(\mathsf{T}K)$$

and

$$s(G)-2s(\mathsf{T}K)-2s(\mathsf{T}C).$$

Then as the system of sets

$$\mathsf{T}\{C+\mathbf{b}_j\} \quad (j = 1, 2, \ldots)$$

covers the whole space, each point of the cube G'' lies in a set

$$\mathsf{T}\{C+\mathbf{b}_i\},$$

which lies completely within G'. Thus if

$$\mathsf{T}\{C+\mathbf{b}_j\} \quad (j = 1, 2, \ldots, M)$$

are the sets of this form that lie completely within G', we have

$$\begin{aligned} M\mu(\mathsf{T}C) &= \sum_{j=1}^{M}\mu(\mathsf{T}\{C+\mathbf{b}_j\}) \\ &\geqslant \mu(G'') = \{s(G)-2s(\mathsf{T}C)-2s(\mathsf{T}K)\}^n. \end{aligned} \tag{17}$$

Since each set $K+\mathbf{a}_i$ has at least one point in C, each of the sets

$$\mathsf{T}\{K+\mathbf{a}_i+\mathbf{b}_j\} \quad (i = 1, 2, \ldots, N;\ j = 1, 2, \ldots, M)$$

has at least one point in G'. Hence each of these sets lies completely in G. Thus

$$\begin{aligned} \rho_{-}(\mathsf{T}\mathscr{K}_P, G) &\geqslant \frac{1}{\mu(G)}\sum_{i=1}^{N}\sum_{j=1}^{M}\mu(\mathsf{T}\{K+\mathbf{a}_i+\mathbf{b}_j\}) \\ &= NM\frac{\mu(\mathsf{T}K)}{\mu(G)}. \end{aligned}$$

So, using (17), we have

$$\begin{aligned} \rho_{-}(\mathsf{T}\mathscr{K}_P, G) &\geqslant N\frac{\mu(\mathsf{T}K)}{\mu(\mathsf{T}C)}\left\{1-\frac{2s(\mathsf{T}C)}{s(G)}-\frac{2s(\mathsf{T}K)}{s(G)}\right\}^n \\ &= N\frac{\mu(K)}{\mu(C)}\left\{1-\frac{2s(\mathsf{T}C)}{s(G)}-\frac{2s(\mathsf{T}K)}{s(G)}\right\}^n. \end{aligned}$$

As this holds for all G with $s(G)$ sufficiently large, it follows that

$$\rho_-(\mathsf{T}\mathscr{K}_P) = \liminf_{s(G)\to\infty} \rho_-(\mathsf{T}\mathscr{K}_P, G) \geqslant \frac{N\mu(K)}{\mu(C)}.$$

A precisely similar argument, taking G' and G'' to be cubes concentric with G with edge-lengths

$$s(G) + 2s(\mathsf{T}K)$$

and

$$s(G) + 2s(\mathsf{T}K) + 2s(\mathsf{T}C),$$

and studying the sets $\mathsf{T}\{K + \mathbf{a}_i + \mathbf{b}_j\}$

that have a point in common with G, and which consequently lie in G', and which so have the property that the corresponding set

$$\mathsf{T}\{C + \mathbf{b}_j\}$$

lies in G'', shows easily enough that

$$\rho_+(\mathsf{T}\mathscr{K}_P) \leqslant \frac{N\mu(K)}{\mu(C)}.$$

Hence

$$\rho_+(\mathsf{T}\mathscr{K}_P) = \rho_-(\mathsf{T}\mathscr{K}_P) = \frac{N\mu(K)}{\mu(C)}.$$

Finally, we obtain the remaining results in (16) by taking T to be the identity.

Theorem 1.6. *Let K be a bounded set with positive measure. Let $\mathscr{K}_L$ be the system of translates of K by the vectors*

$$u_1\mathfrak{g}_1 + u_2\mathfrak{g}_2 + \ldots + u_n\mathfrak{g}_n \quad (u_1, \ldots, u_n = 0, \pm 1, \pm 2, \ldots)$$

of a lattice Λ. Then

$$\rho_+(\mathscr{K}_L) = \rho_-(\mathscr{K}_L) = \frac{\mu(K)}{d(\Lambda)}, \tag{18}$$

where $d(\Lambda)$ is the absolute value of the determinant of the coordinates of the points $\mathfrak{g}_1, \mathfrak{g}_2, \ldots, \mathfrak{g}_n$.

Proof. Let L be the non-singular linear transformation transforming the points

$$\mathbf{e}_1 = (1, 0, \ldots, 0),$$
$$\mathbf{e}_2 = (0, 1, \ldots, 0),$$
$$\ldots\ldots\ldots\ldots\ldots\ldots$$
$$\mathbf{e}_n = (0, 0, \ldots, 1),$$

into the points $\mathfrak{g}_1, \mathfrak{g}_2, \ldots, \mathfrak{g}_n$.

Let C be the cube of edge-length 1 defined by

$$0 \leqslant x_1 < 1, \quad 0 \leqslant x_2 < 1, \ldots, \quad 0 \leqslant x_n < 1.$$

Take $N = 1$ and $\mathbf{a}_1 = \mathbf{o}$. Take $\mathbf{b}_1, \mathbf{b}_2, \ldots$ to be the points that have integers for coordinates. Then the points

$$\mathsf{L}\mathbf{b}_j \quad (j = 1, 2, \ldots)$$

run through the points of the lattice Λ, and the sets

$$\mathsf{L}\{\mathsf{L}^{-1}K + \mathbf{a}_1 + \mathbf{b}_j\} \quad (j = 1, 2, \ldots)$$

are the sets of the system $\mathscr{K}_L$. Hence, by Theorem 1.5 applied to the set $\mathsf{L}^{-1}K$ and the transformation L, we have

$$\rho_+(\mathscr{K}_L) = \rho_-(\mathscr{K}_L) = \frac{\mu(\mathsf{L}^{-1}K)}{\mu(C)} = \frac{\mu(K)}{\mu(\mathsf{L}C)} = \frac{\mu(K)}{d(\Lambda)}.$$

We now prove the main theorem of this section.

THEOREM 1.7. *Let K be a bounded set with a positive measure. Let T be a non-singular affine transformation. Then*

$$\delta(\mathsf{T}K) = \delta_P(K) = \delta(K), \tag{19}$$

$$\delta_L(\mathsf{T}K) = \delta_L(K). \tag{20}$$

Proof. We first prove that

$$\delta(\mathsf{T}K) \geqslant \delta(K). \tag{21}$$

In our proof of this we suppose, as we clearly may, that $\delta(K) > 0$. For each ϵ, with $0 < \epsilon < 1$, we may choose a system $\mathscr{K} = \mathscr{K}_\epsilon$ of translates of K for which

$$\rho_+(\mathscr{K}) > (1-\epsilon)\delta(K).$$

Then we can choose arbitrarily large cubes C for which

$$\rho_+(\mathscr{K}, C) > (1-\epsilon)^2 \delta(K). \tag{22}$$

Take C to be such a cube satisfying also the condition

$$\left\{\frac{s(C)}{s(C) + 2s(K)}\right\}^n > (1-\epsilon). \tag{23}$$

Let C' denote the cube, concentric with C, with edge-length $s(C)+2s(K)$. Then those sets of $\mathscr{K}$ that have a point in common with C lie entirely within C'. Let these sets be

$$K+\mathbf{a}_1,\quad K+\mathbf{a}_2, \ldots, K+\mathbf{a}_N.$$

From the definition of $\rho_+(\mathscr{K}, C)$ and (22), we have

$$\frac{N\mu(K)}{\mu(C)} = \frac{1}{\mu(C)}\sum_{i=1}^{N}\mu(K+\mathbf{a}_i)$$
$$= \rho_+(\mathscr{K}, C) > (1-\epsilon)^2\,\delta(K). \tag{24}$$

Now it is easy to form a packing $\mathscr{C}'$, of translates of the cubes C', which is also a covering of the whole space; it suffices to take the translates of C' by the vectors of the lattice of all points that have coordinates that are integral multiples of $s(C')$. Let $\mathbf{b}_1, \mathbf{b}_2, \ldots$ be an enumeration of these lattice vectors, and consider the periodic system $\mathscr{K}'_P$ of all the sets

$$K+\mathbf{a}_i+\mathbf{b}_j \quad (i = 1, 2, \ldots, N;\ j = 1, 2, \ldots),$$

and the system $\mathsf{T}\mathscr{K}'_P$ of all the sets

$$\mathsf{T}(K+\mathbf{a}_i+\mathbf{b}_j) \quad (i = 1, 2, \ldots, N;\ j = 1, 2, \ldots).$$

It is clear that both systems are packings, and that they are of the type discussed in Theorem 1.5.

By Theorem 1.5 we have

$$\rho_+(\mathsf{T}\mathscr{K}'_P) = \rho_-(\mathsf{T}\mathscr{K}'_P) = \frac{N\mu(K)}{\mu(C')}.$$

Combining this with (23) and (24), we have

$$\rho_+(\mathsf{T}\mathscr{K}'_P) = \rho_-(\mathsf{T}\mathscr{K}'_P) > (1-\epsilon)^2\,\delta(K)\frac{\mu(C)}{\mu(C')} > (1-\epsilon)^3\,\delta(K). \tag{25}$$

Since ϵ may take arbitrarily small values, this implies that

$$\delta(\mathsf{T}K) = \sup_{\mathsf{T}\mathscr{K}} \rho_+(\mathsf{T}\mathscr{K}) \geqslant \delta(K), \tag{26}$$

the supremum being taken over all packings $\mathsf{T}\mathscr{K}$ of $\mathsf{T}K$.

In the case when T is the identity the system $\mathscr{K}'_P$ is periodic, and (25) shows that

$$\delta_P(K) = \sup_{\mathscr{K}_P} \rho_+(\mathscr{K}_P) \geqslant \delta(K). \tag{27}$$

Hence, by Theorem 1.4,

$$\delta_P(K) = \delta(K).$$

Applying the result (26) to the set $\mathsf{T}K$ and the inverse transformation T^{-1}, we obtain

$$\delta(K) = \delta(\mathsf{T}^{-1}\mathsf{T}K) \geqslant \delta(\mathsf{T}K).$$

Thus

$$\delta(\mathsf{T}K) = \delta(K).$$

The proof that $\delta_L(\mathsf{T}K) = \delta_L(K)$ is rather simpler. If $\mathscr{K}_L$ is a lattice packing of K with lattice Λ, then $\mathsf{T}\mathscr{K}_L$ is a lattice packing of $\mathsf{T}K$ with lattice $\mathsf{T}\Lambda - \mathsf{T}\mathbf{o}$. So using Theorem 1.6 twice,

$$\rho_+(\mathsf{T}\mathscr{K}_L) = \rho_-(\mathsf{T}\mathscr{K}_L) = \frac{\mu(\mathsf{T}K)}{d(\mathsf{T}\Lambda - \mathsf{T}\mathbf{o})}$$
$$= \frac{\mu(K)}{d(\Lambda)} = \rho_+(\mathscr{K}_L) = \rho_-(\mathscr{K}_L).$$

Since $\mathsf{T}\mathscr{K}_L$ runs through all lattice packings of $\mathsf{T}K$, as $\mathscr{K}_L$ runs through all lattice packings of K, it follows that

$$\delta_L(\mathsf{T}K) = \delta_L(K).$$

By making the proof of the above theorem slightly more complicated than is strictly necessary, we have arranged to obtain the following theorem as a corollary.

THEOREM 1.8. *If K is a bounded set with positive measure, then*

$$\sup_{\mathscr{K}} \rho_-(\mathscr{K}) = \delta(K), \tag{28}$$

the supremum being taken over all packings $\mathscr{K}$ of K.

Proof. We have trivially

$$\delta(K) = \sup_{\mathscr{K}} \rho_+(\mathscr{K}) \geqslant \sup_{\mathscr{K}} \rho_-(\mathscr{K});$$

the reverse inequality follows from (25), on taking T to be the identity transformation.

3. Invariance of the covering density

By use of the methods developed in the last section, and by making only the modifications suggested by the duality between packings and coverings, it is easy to establish the following theorems, which we state without further proof.

THEOREM 1.9. *Let K be a bounded set with a positive measure. Let T be a non-singular affine transformation. Then*

$$\vartheta(\mathsf{T}K) = \vartheta_P(K) = \vartheta(K), \tag{29}$$

$$\vartheta_L(\mathsf{T}K) = \vartheta_L(K). \tag{30}$$

THEOREM 1.10. *If K is a bounded set with a positive measure, then*

$$\inf_{\mathscr{K}} \rho_+(\mathscr{K}) = \vartheta(K), \tag{31}$$

the infimum being taken over all coverings $\mathscr{K}$ with K.

CHAPTER 2

THE EXISTENCE OF REASONABLY DENSE PACKINGS

1. Packing of convex sets

While we shall be mainly interested in packings of a convex body K, we first prove a more general result. We suppose that K is a bounded open set. We use $\mathsf{D}K$ to denote the *difference body* of K, that is the set of all points expressible vectorially as

$$\mathbf{x}-\mathbf{y},$$

where $\mathbf{x}\in K$ and $\mathbf{y}\in K$. We prove

THEOREM 2.1. *If K is a bounded open set*

$$\delta(K) \geqslant \frac{2\mu(K)}{\mu(\mathsf{D}K)}. \tag{1}$$

Proof. We suppose, as we may, that K contains the origin $\mathbf{o}$ of the coordinate system. Let $s(K)$ denote a number so large that K is contained in a cube with edge-length $s(K)$ (and with its edges parallel to the coordinate axes). Let C be a cube with $s(C) > 2s(K)$. We choose a sequence of points

$$\mathbf{a}_i = (a_i^{(1)}, a_i^{(2)}, \ldots, a_i^{(n)}) \quad (i = 1, 2, 3, \ldots),$$

inductively, to satisfy the following condition. The point $\mathbf{a}_i$ is chosen from among the points $\mathbf{a}$, satisfying the conditions:

(1) $K+\mathbf{a} \subset C$,

(2) $\{K+\mathbf{a}\}\cap\{K+\mathbf{a}_j\} = \varnothing$ for $j < i$,

so that its first coordinate has the least possible value. Note that (2) is vacuous if $i = 1$. Since the conditions (1) and (2) restrict the point $\mathbf{a}$ to a closed bounded set, this choice is always possible until the set of possible points $\mathbf{a}$ becomes empty, and the sequence terminates with some point, $\mathbf{a}_N$ say.

Consider any point $\mathbf{y}$ in the cube C', concentric with C, and with edge-length $s(C)-2s(K)$. Then clearly

$$a_1^{(1)} \leqslant y_1.$$

Let r be the largest integer for which $\mathbf{a}_r$ is defined and

$$a_1^{(r)} \leqslant y_1.$$

Then, either $\mathbf{a}_{r+1}$ is not defined, or

$$a_1^{(r+1)} > y_1.$$

In either case it follows that $\mathbf{y}$ was not a possible choice for a point $\mathbf{a}_{r+1}$. Hence, either

(1) $K+\mathbf{y}$ is not contained in C, or

(2) $K+\mathbf{y}$ has a point in common with one of the sets $K+\mathbf{a}_j$ $(j = 1, 2, \ldots, r)$.

The first possibility is excluded as $\mathbf{y}$ belongs to C'. Hence there are points $\mathbf{k}_1$ and $\mathbf{k}_2$ in K, and an integer j with $1 \leqslant j \leqslant r$, such that

$$\mathbf{k}_1 + \mathbf{a}_j = \mathbf{k}_2 + \mathbf{y}.$$

Hence

$$\mathbf{y} = \mathbf{k}_1 - \mathbf{k}_2 + \mathbf{a}_j;$$

and so $\mathbf{y}$ belongs to

$$\mathsf{D}K + \mathbf{a}_j.$$

Since

$$y_1 \geqslant a_1^{(j)}$$

it follows that $\mathbf{y}$ belongs to

$$\mathsf{HD}K + \mathbf{a},$$

where $\mathsf{HD}K$ denotes the set of points $\mathbf{x}$ of $\mathsf{D}K$ with $x_1 \geqslant 0$. This shows that the sets

$$\mathsf{HD}K + \mathbf{a}_i \quad (i = 1, 2, \ldots, N)$$

cover the cube C'. We may suppose that

$$\mathsf{HD}K + \mathbf{a}_i \quad (i = 1, 2, \ldots, N')$$

are those of the sets that have a point in common with C'.

Now let $\mathbf{b}_1$, $\mathbf{b}_2$ ... be the points of the lattice of all points that have integral multiples of $s(C)$ for coordinates, and let $\mathbf{b}_1'$, $\mathbf{b}_2'$ ... be the points of the lattice of all points that have integral multiples of $s(C')$ for coordinates. Then the system $\mathscr{K}_P$ of sets

$$K + \mathbf{a}_i + \mathbf{b}_j \quad (i = 1, 2, \ldots, N;\ j = 1, 2, \ldots)$$

forms a packing of K, and the system $\mathscr{H}_P$ of sets

$$\mathsf{HD}K + \mathbf{a}_i + \mathbf{b}_j' \quad (i = 1, 2, \ldots, N';\ j = 1, 2, \ldots)$$

forms a covering with $H = \mathsf{HD}K$.

By Theorem 1.5 we have

$$\rho_+(\mathscr{K}_P) = \frac{N\mu(K)}{\mu(C)}$$

and

$$\rho_-(\mathscr{H}_P) = \frac{N'\mu(\mathsf{HD}K)}{\mu(C')}$$

$$= \frac{N'\mu(\mathsf{HD}K)}{\mu(C)}\left\{1-\frac{2s(K)}{s(C)}\right\}^{-n}.$$

But since $\mathscr{K}_P$ is a packing and $\mathscr{H}_P$ is a covering

$$\delta(K) \geqslant \rho_+(\mathscr{K}_P), \quad \vartheta(H) \leqslant \rho_-(\mathscr{H}_P).$$

Combining these results

$$\delta(K) \geqslant \frac{N\mu(K)}{\mu(C)} \geqslant \frac{\mu(K)}{\mu(\mathsf{HD}K)}\,\frac{N'\mu(\mathsf{HD}K)}{\mu(C)}$$

$$\geqslant \frac{\mu(K)}{\mu(\mathsf{HD}K)}\left\{1-\frac{2s(K)}{s(C)}\right\}^{n}\vartheta(\mathsf{HD}K).$$

Since $s(C)$ may be taken arbitrarily large, this implies that

$$\delta(K) \geqslant \frac{\mu(K)}{\mu(\mathsf{HD}K)}\vartheta(\mathsf{HD}K). \tag{2}$$

The required inequality (1) now follows, on noting that $\mathsf{D}K$ is symmetrical in $\mathbf{o}$, so that

$$\mu(\mathsf{HD}K) = \tfrac{1}{2}\mu(\mathsf{D}K)$$

and that

$$\vartheta(\mathsf{HD}K) \geqslant 1.$$

We note that any covering of space with the sets $\mathsf{HD}K$ automatically leads to a corresponding covering, with twice the density, by the same translations applied to the set $\mathsf{D}K$. Hence

$$\vartheta(\mathsf{HD}K) \geqslant \tfrac{1}{2}\vartheta(\mathsf{D}K).$$

So we have as a corollary:

Corollary.

$$\delta(K) \geqslant \frac{\mu(K)}{\mu(\mathsf{D}K)}\vartheta(\mathsf{D}K). \tag{3}$$

In the case when K is a convex body which is symmetrical

in **o**, the set $\mathsf{D}K$ reduces to the set $2K$ of all points of the form $2\mathbf{k}$ with $\mathbf{k}$ in K. Thus

$$\mu(\mathsf{D}K) = \mu(2K) = 2^n\mu(K),$$

and we obtain the following theorem.

THEOREM 2.2. *If K is an open convex body with* **o** *as centre*

$$\delta(K) \geqslant (\tfrac{1}{2})^{n-1} \tag{4}$$

and

$$\delta(K) \geqslant (\tfrac{1}{2})^n \vartheta(K). \tag{5}$$

In the next section we show that, for any convex body K,

$$\mu(\mathsf{D}K) \leqslant \binom{2n}{n}\mu(K).$$

Consequently, combining this with Theorem 2.1, we shall have

THEOREM 2.3. *If K is an open convex body*

$$\delta(K) \geqslant \frac{2(n!)^2}{(2n)!}. \tag{6}$$

In these three theorems the restriction to open sets is easily removed; it suffices merely to know that K has positive measure. Refinements will be discussed in Chapter 4; the limitations on their possible improvements will be discussed in Chapters 6 and 7.

In most cases nothing significant is known about the factor $\vartheta(\mathsf{D}K)$ occurring in (3), so that the theorem is more powerful than its corollary; similarly (4) is more powerful than (5). An exception is the case when K is an n-dimensional sphere. In this case by appealing to Theorem 8.1 of Chapter 8, which is independent of the present work, we obtain

THEOREM 2.4. *If K is a sphere, then*

$$\delta(K) \geqslant (\tfrac{1}{2})^n \tau_n, \tag{7}$$

where τ_n is the ratio defined in Chapter 8.

Here, by (15) of Chapter 8,

$$\tau_n \sim \frac{n}{e\sqrt{e}} \quad (n \to \infty).$$

2. The volume of the difference set

In this section we prove the following theorem used in the last section.

THEOREM 2.4. *If K is a convex set, then*

$$\mu(\mathsf{D}K) \leqslant \binom{2n}{n} \mu(K), \tag{8}$$

with equality when K is a simplex.

Proof. We suppose, as we may, that K is closed. Let $\chi(x)$ be the characteristic function of K, that is the function taking the value 1 at all points in K, and taking the value 0 at all points not in K. We first evaluate a simple integral, by changing the order of integration. We have

$$\begin{aligned} \int\left[\int \chi(\mathbf{y}-\mathbf{x})\chi(\mathbf{y})\,d\mathbf{y}\right]d\mathbf{x} &= \int \chi(\mathbf{y})\left[\int \chi(\mathbf{y}-\mathbf{x})\,d\mathbf{x}\right]d\mathbf{y} \\ &= \int \chi(\mathbf{y})\mu(K)\,d\mathbf{y} = [\mu(K)]^2, \end{aligned} \tag{9}$$

the integrals being taken over the whole of space.

We note that for each $\mathbf{x}$ the integral

$$\int \chi(\mathbf{y}-\mathbf{x})\chi(\mathbf{y})\,d\mathbf{y}$$

will be zero, unless there is some point $\mathbf{y}$, such that $\mathbf{y}$ and $\mathbf{y}-\mathbf{x}$ both belong to K. But in this case the point

$$\mathbf{x} = \mathbf{y} - (\mathbf{y}-\mathbf{x})$$

belongs to $\mathsf{D}K$. Hence the equation (9) can be written in the form

$$\int_{\mathsf{D}K}\left[\int \chi(\mathbf{y}-\mathbf{x})\chi(\mathbf{y})\,d\mathbf{y}\right]d\mathbf{x} = [\mu(K)]^2. \tag{10}$$

For each point $\mathbf{x}$ of $\mathsf{D}K$, other than $\mathbf{o}$, we introduce a scalar $\lambda = \lambda(\mathbf{x})$ with $0 < \lambda \leqslant 1$, so that the point $\lambda^{-1}\mathbf{x}$ lies on the boundary of $\mathsf{D}K$. We write

$$\mathbf{z} = \mathbf{z}(\mathbf{x}) = \lambda^{-1}\mathbf{x}.$$

Since K is closed and bounded so is $\mathsf{D}K$, and $\mathbf{z}$ lies in $\mathsf{D}K$. So there are points $\mathbf{a}$, $\mathbf{b}$ in K with

$$\mathbf{z} = \mathbf{b} - \mathbf{a}.$$

Now, by convexity, the set

$$(1-\lambda)K + \lambda\mathbf{b}$$

is contained in K, and the set

$$(1-\lambda)K + \lambda\mathbf{a} + \mathbf{x}$$

is contained in $K + \mathbf{x}$. But, since

$$\lambda\mathbf{b} - (\lambda\mathbf{a} + \mathbf{x}) = \lambda(\mathbf{b} - \mathbf{a}) - \mathbf{x} = \lambda\mathbf{z} - \mathbf{x} = \mathbf{o},$$

the two sets coincide, and so each is contained in

$$K \cap \{K + \mathbf{x}\}.$$

Thus
$$\begin{aligned}\int \chi(\mathbf{y}-\mathbf{x})\chi(\mathbf{y})\,d\mathbf{y} &= \mu(K \cap \{K+\mathbf{x}\}) \\ &\geqslant \mu(\{1-\lambda\}K) \\ &= \{1-\lambda\}^n \mu(K) = \{1-\lambda(\mathbf{x})\}^n \mu(K). \qquad (11)\end{aligned}$$

Substituting this into (10), we obtain

$$[\mu(K)]^2 \geqslant \int_{\mathsf{D}K} \{1-\lambda(\mathbf{x})\}^n \mu(K)\,d\mathbf{x}.$$

Now we divide by $\mu(K)$ and write

$$\{1-\lambda(\mathbf{x})\}^n = \int_{\lambda(\mathbf{x})}^1 n(1-t)^{n-1}\,dt.$$

Then
$$\begin{aligned}\mu(K) &\geqslant \int_{\mathsf{D}K} \left[\int_{\lambda(\mathbf{x})}^1 n(1-t)^{n-1}\,dt\right] d\mathbf{x} \\ &= \int_0^1 \left[\int_{\substack{\mathbf{x}\in\mathsf{D}K \\ \lambda(\mathbf{x})\leqslant t}} n(1-t)^{n-1}\,d\mathbf{x}\right] dt. \qquad (12)\end{aligned}$$

But, since the point $(\lambda(\mathbf{x}))^{-1}\mathbf{x}$ was adjusted to lie on the boundary of $\mathsf{D}K$, the point $\mathbf{x}$ lies on the boundary of $\lambda(\mathbf{x})\,\mathsf{D}K$, and the condition $\lambda(\mathbf{x}) \leqslant t$ is equivalent to the condition

$$\mathbf{x} \in t\mathsf{D}K.$$

Consequently
$$\int_0^1 \left[\int_{\substack{\mathbf{x} \in \mathsf{D}K \\ \lambda(\mathbf{x}) \leqslant t}} n(1-t)^{n-1} d\mathbf{x} \right] dt$$
$$= \int_0^1 n(1-t)^{n-1} \mu(t\mathsf{D}K)\, dt$$
$$= \mu(\mathsf{D}K) \int_0^1 n(1-t)^{n-1} t^n\, dt$$
$$= \frac{(n!)^2}{(2n)!} \mu(\mathsf{D}K).$$

Substituting this back into (12) we obtain the required inequality
$$\mu(\mathsf{D}K) \leqslant \frac{(2n)!}{(n!)^2} \mu(K).$$

In the special case when K is a simplex, it is easy to verify that the sets
$$(1-\lambda)\, K + \lambda \mathbf{b},$$
$$(1-\lambda) K + \lambda \mathbf{a} + \mathbf{x},$$
$$K \cap \{K + \mathbf{x}\},$$
all coincide. Thus (11) and all the subsequent inequalities become equalities.

For an alternative proof that (8) holds with equality, when K is a simplex, see Theorem 6.2. For a proof that equality holds in (8) only when K is a simplex, see Rogers and Shephard (1957), or a note that B. Grünbaum has written and will I hope publish.

CHAPTER 3

THE EXISTENCE OF REASONABLY ECONOMICAL COVERINGS

1. Covering most of space

Let K be a bounded set with positive measure, and let $\mathscr{K}$ be the system $\{K+\mathbf{a}_i\}$ of translates of K by the vectors $\mathbf{a}_1, \mathbf{a}_2, \ldots$. In general the system $\mathscr{K}$ will not cover the whole of space, and the question arises: 'What proportion of space is covered by the sets of $\mathscr{K}$?' We introduce a measure of the proportion of space which is left uncovered, in the following way. For each cube C, with edge-length $s(C)$, write

$$\epsilon(\mathscr{K}, C) = \frac{1}{\mu(C)}\left[\mu(C) - \mu\left(\bigcup_{i=1}^{\infty} \{(K+\mathbf{a}_i)\cap C\}\right)\right],$$

so that $\epsilon(\mathscr{K}, C)$ is the proportion of the cube C which is left uncovered by the sets of the system $\mathscr{K}$. We then write

$$\epsilon_+(\mathscr{K}) = \limsup_{s(C)\to\infty} \epsilon(\mathscr{K}, C), \tag{1}$$

so that it is safe to say that if $\epsilon_+(\mathscr{K})$ is small then the sets of $\mathscr{K}$ cover most of space.

We now prove

THEOREM 3.1. *Let K be a bounded set with positive measure, and let ρ be any positive number. Then there is a system $\mathscr{K}$ of translates of K, with*

$$\rho_+(\mathscr{K}) = \rho_-(\mathscr{K}) = \rho \tag{2}$$

and

$$\epsilon_+(\mathscr{K}) < e^{-\rho}. \tag{3}$$

Proof. Suppose that K is contained in a cube of side $s(K)$. Let C be the cube

$$0 \leqslant x_1 < s(C),\quad 0 \leqslant x_2 < s(C), \ldots,\quad 0 \leqslant x_n < s(C)$$

of side $s(C)$, with $s(C) > 2s(K)$, chosen so that the number

$$\{s(C)\}^n \rho/\mu(K)$$

has an integral value, say N.

Let Λ denote the lattice of all points whose coordinates are integral multiples of $s(C)$, and let $\mathbf{b}_1, \mathbf{b}_2, \ldots$ be an enumeration of the points of Λ. Let $\mathbf{x}_1, \mathbf{x}_2, \ldots, \mathbf{x}_N$ be a system of N points lying in C, and consider the system $\mathscr{K} = \mathscr{K}(\mathbf{x}_1, \mathbf{x}_2, \ldots, \mathbf{x}_N)$ of the translates

$$K + \mathbf{x}_i + \mathbf{b}_j \quad (i = 1, 2, \ldots, N; j = 1, 2, \ldots).$$

By Theorem 1.5

$$\rho_+(\mathscr{K}) = \rho_-(\mathscr{K}) = \frac{N\mu(K)}{\{s(C)\}^n} = \rho$$

as required.

Since $s(C) > 2s(K)$, two sets

$$K + \mathbf{x}_i + \mathbf{b}_j, \quad K + \mathbf{x}_i + \mathbf{b}_k$$

can only have a point in common if $j = k$ and they coincide. Hence, if $\chi(\mathbf{x})$ is the characteristic function of K, the characteristic function of the set

$$\bigcup_{j=1}^{\infty} \{K + \mathbf{x}_i + \mathbf{b}_j\}$$

is

$$\sum_{j=1}^{\infty} \chi(\mathbf{x} - \mathbf{x}_i - \mathbf{b}_j)$$

for $i = 1, 2, \ldots, N$. So the characteristic function of the set E of points not covered by any set of $\mathscr{K}$ is

$$\epsilon(\mathbf{x}) = \prod_{i=1}^{N} \left[1 - \sum_{j=1}^{\infty} \chi(\mathbf{x} - \mathbf{x}_i - \mathbf{b}_j)\right]. \tag{4}$$

Thus, if G is any cube (with its edges parallel to the coordinate axes), we have

$$\epsilon(\mathscr{K}, G) = \frac{1}{\mu(G)} \int_G \epsilon(\mathbf{x})\, d\mathbf{x}.$$

But the integrand is periodic, with period $s(C)$, in each of the coordinates. It follows easily that

$$\begin{aligned} \epsilon_+(\mathscr{K}) &= \limsup_{s(G) \to +\infty} \epsilon(\mathscr{K}, G) \\ &= \frac{1}{\mu(C)} \int_C \epsilon(\mathbf{x})\, d\mathbf{x}. \end{aligned} \tag{5}$$

Now, to prove the existence of a system $\mathscr{K}$ satisfying (2) and (3), it suffices to prove that the average value of

$$\epsilon_+\{\mathscr{K}(\mathbf{x}_1, \mathbf{x}_2, \ldots, \mathbf{x}_N)\},$$

taken over all choices of $\mathbf{x}_1, \mathbf{x}_2, \ldots, \mathbf{x}_N$ in C, is less than $e^{-\rho}$. But this average value is

$$\frac{1}{[\mu(C)]^N}\int_C\int_C\cdots\int_C \epsilon_+\{\mathscr{K}(\mathbf{x}_1, \mathbf{x}_2, \ldots, \mathbf{x}_N)\}\,d\mathbf{x}_1 d\mathbf{x}_2 \ldots d\mathbf{x}_N$$

$$= \frac{1}{[\mu(C)]^{N+1}}\int_C\int_C\cdots\int_C\left\{\int_C \epsilon(\mathbf{x})\,d\mathbf{x}\right\} d\mathbf{x}_1 d\mathbf{x}_2 \ldots d\mathbf{x}_N$$

$$= \frac{1}{[\mu(C)]^{N+1}}\int_C\int_C\cdots\int_C$$

$$\times\left\{\int_C \prod_{i=1}^{N}\left[1-\sum_{j=1}^{\infty}\chi(\mathbf{x}-\mathbf{x}_i-\mathbf{b}_j)\right]d\mathbf{x}\right\} d\mathbf{x}_1 d\mathbf{x}_2 \ldots d\mathbf{x}_N.$$

By successively changing the order of integration, factorizing the inner integral, changing the order of integration and summation, and making the substitutions

$$\mathbf{x}_i = \mathbf{y}_i + \mathbf{x} - \mathbf{b}_j,$$

we can express the average value in the forms

$$\frac{1}{[\mu(C)]^{N+1}}\int_C\left\{\int_C\int_C\cdots\int_C\right.$$

$$\left.\times\prod_{i=1}^{N}\left[1-\sum_{j=1}^{\infty}\chi(\mathbf{x}-\mathbf{x}_i-\mathbf{b}_j)\right]d\mathbf{x}_1 d\mathbf{x}_2 \ldots d\mathbf{x}_N\right\} d\mathbf{x}$$

$$= \frac{1}{[\mu(C)]^{N+1}}\int_C \prod_{i=1}^{N}\left\{\int_C\left[1-\sum_{j=1}^{\infty}\chi(\mathbf{x}-\mathbf{x}_i-\mathbf{b}_j)\right]d\mathbf{x}_i\right\} d\mathbf{x}$$

$$= \frac{1}{[\mu(C)]^{N+1}}\int_C \prod_{i=1}^{N}\left\{\mu(C)-\sum_{j=1}^{\infty}\int_C \chi(\mathbf{x}-\mathbf{x}_i-\mathbf{b}_j)\,d\mathbf{x}_i\right\} d\mathbf{x}$$

$$= \frac{1}{[\mu(C)]^{N+1}}\int_C \prod_{i=1}^{N}\left\{\mu(C)-\sum_{j=1}^{\infty}\int_{C-\mathbf{x}+\mathbf{b}_j} \chi(-\mathbf{y}_i)\,d\mathbf{y}_i\right\} d\mathbf{x}.$$

But the sets $\quad C-\mathbf{x}+\mathbf{b}_j \quad (j = 1, 2, \ldots)$

fit together exactly to make up the whole of space. So the average value reduces to

$$\frac{1}{[\mu(C)]^{N+1}}\int_C \prod_{i=1}^{N}\left\{\mu(C)-\int \chi(-\mathbf{y})\,d\mathbf{y}\right\}d\mathbf{x}$$

$$=\frac{1}{[\mu(C)]^{N+1}}\int_C \prod_{i=1}^{N}\{\mu(C)-\mu(K)\}\,d\mathbf{x}$$

$$=\left(1-\frac{\mu(K)}{\mu(C)}\right)^N=\left(1-\frac{\rho}{N}\right)^N<e^{-\rho}.$$

This completes the proof.

2. Covering the whole of space

Although the last section shows that it is possible to obtain reasonably economical coverings of most of space, we need to make further assumptions, concerning the set K, before we can obtain coverings of the whole of space. Although more general results can be obtained by the method we use, we confine our attention to the case when K is convex and prove

THEOREM 3.2. *If K is a convex body, then*

$$\vartheta(K)\leqslant n\log n+n\log\log n+5n, \tag{6}$$

provided $n \geqslant 2$.

Proof. We may suppose, without loss of generality, that the origin $\mathbf{o}$ coincides with the centroid of K. Then by a well-known result in the theory of convex bodies (see Bonnesen and Fenchel, 1934, pp. 52–3), the set $-n^{-1}K$ is contained in K, i.e.

$$-n^{-1}K\subset K. \tag{7}$$

Now we suppose that the construction used in the proof of Theorem 3.1 is carried out again. A cube of side $s(K)$ is chosen containing K. A cube C of the form

$$0\leqslant x_1<s(C),\quad 0\leqslant x_2<s(C),\ldots,0\leqslant x_n<s(C)$$

is chosen so that
$$s(C)>2s(K)$$
and
$$N=\{s(C)\}^n\rho/\mu(K) \tag{8}$$

is integral. Points $\mathbf{b}_1, \mathbf{b}_2, \ldots$ are chosen so that the sets

$$C+\mathbf{b}_j \quad (j = 1, 2, \ldots)$$

fit together in a lattice arrangement to make up the whole space. Then $\mathscr{K}$ is taken to be the system of sets

$$K+\mathbf{a}_i+\mathbf{b}_j \quad (i = 1, 2, \ldots, N;\, j = 1, 2, \ldots),$$

where $\mathbf{a}_1, \mathbf{a}_2, \ldots, \mathbf{a}_N$ are points in C specially chosen to ensure that

$$\epsilon_+(\mathscr{K}) < e^{-\rho}.$$

We note that the construction ensures that

$$\epsilon_+(\mathscr{K}) = \frac{1}{\mu(C)}\int_C \epsilon(\mathbf{x})\,d\mathbf{x} = \frac{\mu(C \cap E)}{\mu(C)},$$

$\epsilon(\mathbf{x})$ being the characteristic function of the set E of points not covered by the sets of the system $\mathscr{K}$.

Let η be any real number satisfying $0 < \eta < 1/n$. It may happen that there are points $\mathbf{c}_1, \mathbf{c}_2, \ldots, \mathbf{c}_M$ of C such that the bodies

$$-\eta K+\mathbf{b}_j+\mathbf{c}_k \quad (j = 1, 2, \ldots;\, k = 1, 2, \ldots, M) \tag{9}$$

do not have any points in common with each other nor with any set of the system $\mathscr{K}$. For a fixed value of k the bodies

$$-\eta K+\mathbf{b}_j+\mathbf{c}_k \quad (j = 1, 2, \ldots)$$

form a packing into the whole space, and, by the periodicity of the system, we have

$$\mu\left(C \cap \bigcup_{j=1}^{\infty}\{-\eta K+\mathbf{b}_j+\mathbf{c}_k\}\right) = \mu(-\eta K+\mathbf{c}_k) = \mu(-\eta K) = \eta^n\mu(K).$$

Since the sets

$$\bigcup_{j=1}^{\infty}\{-\eta K+\mathbf{b}_j+\mathbf{c}_k\} \quad (k = 1, 2, \ldots, M)$$

are all disjoint, and all lie in E, we deduce that

$$\begin{aligned}\mu(C \cap E) &\geqslant \mu\left(C \cap \bigcup_{k=1}^{M}\bigcup_{j=1}^{\infty}\{-\eta K+\mathbf{b}_j+\mathbf{c}_k\}\right)\\ &= \sum_{k=1}^{M}\mu\left(C \cap \bigcup_{j=1}^{\infty}\{-\eta K+\mathbf{b}_j+\mathbf{c}_k\}\right) = M\eta^n\mu(K).\end{aligned}$$

Hence

$$M \leqslant \eta^{-n}\frac{\mu(C \cap E)}{\mu(K)} = \eta^{-n}\frac{\mu(C)}{\mu(K)}\frac{\mu(C \cap E)}{\mu(C)}$$

$$= \eta^{-n}\frac{\mu(C)}{\mu(K)}\epsilon_+(\mathscr{K}) < \eta^{-n}\frac{\mu(C)}{\mu(K)}e^{-\rho}.$$

If there is no such system of points $\mathbf{c}_1, \mathbf{c}_2, \ldots, \mathbf{c}_M$ we take $M = 0$; otherwise we take M to have its largest possible value, and choose a corresponding system of points $\mathbf{c}_1, \mathbf{c}_2, \ldots, \mathbf{c}_M$. In either case we have

$$M \leqslant \eta^{-n}\frac{\mu(C)}{\mu(K)}e^{-\rho}. \tag{10}$$

Now we study a general point $\mathbf{x}$ of C, and prove that it belongs to at least one of a certain system of convex bodies. Since $0 < \eta < 1$ it follows that no two of the bodies

$$-\eta K + \mathbf{b}_j + \mathbf{x} \quad (j = 1, 2, \ldots)$$

have any point in common. Hence, by the choice of M to have its largest possible value, at least one of the bodies of this system, say the body

$$-\eta K + \mathbf{b}_k + \mathbf{x}, \tag{11}$$

has a point in common with one of the bodies of the system (9) or of the system $\mathscr{K}$. First, suppose that there is a point common to the body (11) and the body

$$K + \mathbf{a}_i + \mathbf{b}_j,$$

say, of the system $\mathscr{K}$. Then

$$-\eta\mathbf{k}_1 + \mathbf{b}_k + \mathbf{x} = \mathbf{k}_2 + \mathbf{a}_i + \mathbf{b}_j$$

for some points $\mathbf{k}_1$ and $\mathbf{k}_2$ in K. Thus

$$\mathbf{x} = \mathbf{k}_2 + \eta\mathbf{k}_1 + \mathbf{a}_i + \mathbf{b}_j - \mathbf{b}_k.$$

Since K is convex $\quad \mathbf{k}_2 + \eta\mathbf{k}_1 \in (1+\eta)K,$

and since $\mathbf{b}_1, \mathbf{b}_2, \ldots$ run through the points of a lattice

$$\mathbf{b}_j - \mathbf{b}_k = \mathbf{b}_l,$$

for some positive integer l. Consequently, $\mathbf{x}$ belongs to

$$(1+\eta)K + \mathbf{a}_i + \mathbf{b}_l.$$

Now suppose that there is a point common to the body

$$-\eta K+\mathbf{b}_k+\mathbf{x}$$

and the body $$-\eta K+\mathbf{b}_j+\mathbf{c}_i$$

of the system (9). Then

$$-\eta\mathbf{k}_1+\mathbf{b}_k+\mathbf{x} = -\eta\mathbf{k}_2+\mathbf{b}_j+\mathbf{c}_i$$

for some points $\mathbf{k}_1$ and $\mathbf{k}_2$ in K. Thus

$$\mathbf{x} = -\eta\mathbf{k}_2+\eta\mathbf{k}_1+(\mathbf{b}_j-\mathbf{b}_k)+\mathbf{c}_i.$$

But by (7) we can write $$-\eta\mathbf{k}_2 = \mathbf{k}_3$$

for some $\mathbf{k}_3$ in K. So, arguing as before, $\mathbf{x}$ belongs to

$$(1+\eta)K+\mathbf{b}_l+\mathbf{c}_i$$

for some positive integer l.

This shows that the whole of C is covered by the system $\mathscr{H}$ of bodies:

$$(1+\eta)K+\mathbf{a}_i+\mathbf{b}_j \quad (i = 1,2,\ldots,N;\ j = 1,2,\ldots),$$

$$(1+\eta)K+\mathbf{c}_k+\mathbf{b}_j \quad (k = 1,2,\ldots,M;\ j = 1,2,\ldots).$$

Using the periodicity of the system, we easily see that the whole of space is covered by $\mathscr{H}$.

Now, by Theorem 1.5,

$$\rho_+(\mathscr{H}) = \rho_-(\mathscr{H}) = \frac{(N+M)\mu[(1+\eta)K]}{\mu(C)}.$$

So, writing $H = (1+\eta)K$, we have

$$\vartheta(H) \leqslant (N+M)(1+\eta)^n\frac{\mu(K)}{\mu(C)}.$$

Thus, by Theorem 1.9, we have

$$\vartheta(K) = \vartheta(H) \leqslant (N+M)(1+\eta)^n\frac{\mu(K)}{\mu(C)}.$$

Hence, by the estimates (8) and (10) for N and M,

$$\vartheta(K) \leqslant [\rho+\eta^{-n}e^{-\rho}](1+\eta)^n.$$

Here ρ is at our disposal; to obtain the best result we take $\rho = n \log(1/\eta)$ to give

$$\vartheta(K) \leqslant \min_{0<\eta<1/n} (1+\eta)^n \{1 + n \log(1/\eta)\}.$$

Provided $n \geqslant 3$ we can take $\eta = 1/(n \log n)$ and obtain

$$\begin{aligned}\vartheta(K) &\leqslant (1+\eta)^n \{1 + n \log(1/\eta)\} \\ &< e^{n\eta}\{1 + n \log(1/\eta)\} \\ &= e^{1/(\log n)}\{1 + n \log(n \log n)\} \\ &< \left\{1 + \frac{2}{\log n}\right\}\{n \log n + n \log\log n + 1\} \\ &< n \log n + n \log\log n + 5n.\end{aligned}$$

When $n = 2$ we easily obtain the same inequality by taking $\eta = 1/e$.

CHAPTER 4

THE EXISTENCE OF REASONABLY DENSE LATTICE PACKINGS

1. Averages over certain sets of lattices

If $\mathbf{a}_1, \mathbf{a}_2, \ldots, \mathbf{a}_n$ are n linearly independent vectors in n-dimensional space, we use the symbol $\Lambda(\mathbf{a}_1, \mathbf{a}_2, \ldots, \mathbf{a}_n)$, or simply Λ, to denote the set of all points of the form

$$\mathbf{a} = u_1\mathbf{a}_1 + u_2\mathbf{a}_2 + \ldots + u_n\mathbf{a}_n,$$

where $u_1, u_2, \ldots, u_n$ are arbitrary integers. We call any such set a lattice. In particular, if

$$\begin{aligned} \mathbf{e}_1 &= (1, 0, \ldots, 0), \\ \mathbf{e}_2 &= (0, 1, \ldots, 0), \\ &\ldots\ldots\ldots\ldots\ldots\ldots \\ \mathbf{e}_n &= (0, 0, \ldots, 1), \end{aligned}$$

the corresponding lattice

$$\Lambda_0 = \Lambda(\mathbf{e}_1, \mathbf{e}_2, \ldots, \mathbf{e}_n)$$

is simply the set of points with integral coordinates. Furthermore, in general, we have

$$\Lambda(\mathbf{a}_1, \mathbf{a}_2, \ldots, \mathbf{a}_n) = \mathsf{T}\Lambda_0,$$

where T is the non-singular linear transformation with its matrix, having the column vectors $\mathbf{a}_1, \mathbf{a}_2, \ldots, \mathbf{a}_n$ as its columns.

While the main interest of the theory of lattices arises through their connections with the theory of numbers, our main reason for introducing them is that their points are distributed throughout the whole space in a very regular way.

We shall confine our attention to very special subsets of the set of all lattices. For all $\eta \geqslant 0$, and all real numbers

$$\alpha_1, \alpha_2, \ldots, \alpha_{n-1},$$

let

$$\Lambda(\alpha_1, \alpha_2, \ldots, \alpha_{n-1}; \eta)$$

denote the lattice $\Lambda(\mathbf{a}_1, \mathbf{a}_2, \ldots, \mathbf{a}_n)$, generated by the points

$$\left.\begin{aligned} \mathbf{a}_1 &= (\chi, 0, 0, \ldots, 0, 0), \\ \mathbf{a}_2 &= (0, \chi, 0, \ldots, 0, 0), \\ &\cdots\cdots\cdots\cdots\cdots\cdots \\ \mathbf{a}_{n-1} &= (0, 0, 0, \ldots, \chi, 0), \\ \mathbf{a}_n &= (\alpha_1\chi, \alpha_2\chi, \alpha_3\chi, \ldots, \alpha_{n-1}\chi, \eta), \end{aligned}\right\} \tag{1}$$

where $\chi = \eta^{-1/(n-1)}$. Let $\Lambda(\eta)$ denote the set of all lattices $\Lambda(\alpha_1, \alpha_2, \ldots, \alpha_{n-1}; \eta)$ with

$$0 \leqslant \alpha_1 < 1, \quad 0 \leqslant \alpha_2 < 1, \ldots, 0 \leqslant \alpha_{n-1} < 1.$$

Note that the determinant of the coordinates of the points $\mathbf{a}_1, \mathbf{a}_2, \ldots, \mathbf{a}_n$ is $\chi^{n-1}\eta = 1$; thus the lattice is obtained by applying a volume preserving linear transformation to the lattice of points with integral coordinates.

It may seem that the lattice $\Lambda(\alpha_1, \ldots, \alpha_{n-1}, \eta)$, depending on n parameters, is necessarily of a very special type. However, it has become clear† that every lattice with determinant 1 can be approximated, arbitrarily closely, by taking η to be small and choosing $\alpha_1, \alpha_2, \ldots, \alpha_{n-1}$ suitably.

Before we state and prove our main lemma, it is convenient to introduce a generalized limit process, and to establish one of its properties. We say that a function $F(\eta)$, defined for $\eta > 0$, tends to a limit F in mean as η tends to 0 from above, and we write

$$\operatorname*{l.i.m.}_{\eta\to+0} F(\eta) = F,$$

provided

$$\lim_{\eta\to+0} \frac{1}{\eta}\int_0^\eta F(\zeta)\,d\zeta = F. \tag{2}$$

LEMMA. *If $f(x)$ is integrable in the Lebesgue sense over $(-\infty, +\infty)$*

$$\operatorname*{l.i.m.}_{\eta\to+0} \sum_{\substack{n=-\infty \\ n\neq 0}}^{\infty} f(n\eta)\eta = \int_{-\infty}^{\infty} f(x)\,dx. \tag{3}$$

Proof. By changing the order of integration and summation, then writing

$$n\zeta = \xi, \quad m = |n|,$$

† To the author.

and finally changing the order of summation and integration back we formally obtain

$$\frac{1}{\eta}\int_0^{\eta} \sum_{\substack{n=-\infty \\ n\neq 0}}^{\infty} f(n\zeta)\,\zeta\,d\zeta$$

$$= \sum_{\substack{n=-\infty \\ n\neq 0}}^{\infty} \int_0^{\eta} (\zeta/\eta) f(n\zeta)\,d\zeta$$

$$= \sum_{m=1}^{\infty} \int_{-m\eta}^{m\eta} (|\xi|/\eta)\, m^{-2} f(\xi)\,d\xi$$

$$= \int_{-\infty}^{\infty} \left(\sum_{m>|\xi|/\eta} \frac{|\xi|}{m^2\eta} \right) f(\xi)\,d\xi. \tag{4}$$

Since

$$\sum_{m>|\xi|/\eta} \frac{|\xi|}{m^2\eta} \leqslant \sum_{m>|\xi|/\eta} \frac{2|\xi|}{m(m+1)\eta} = \frac{2|\xi|}{\eta} \Big/ \left\{ \left[\frac{|\xi|}{\eta}\right] + 1 \right\} < 2, \tag{5}$$

it is clear that the right-hand side of (4) is absolutely convergent, and working backwards the formal result is easily justified. It also follows from (5) that the convergence of

$$\sum_{m>|\xi|/\eta} \frac{|\xi|}{m^2\eta}$$

to its limit 1, as $\eta \to +0$, is bounded. Hence, by the theory of dominated convergence,

$$\lim_{n\to+0} \int_{-\infty}^{\infty} \left(\sum_{m>|\xi|/\eta} \frac{|\xi|}{m^2\eta} \right) f(\xi)\,d\xi = \int_{-\infty}^{\infty} f(x)\,dx.$$

The required result (3) follows from (4) and the definition (2).

We are now in a position to state our main lemma.

THEOREM 4.1. *Provided the integral on the right exists (as a Lebesgue integral), and $f(\mathbf{x})$ vanishes outside a bounded region,*

$$\operatorname*{l.i.m.}_{\eta\to+0} \int_0^1 \int_0^1 \dots \int_0^1 \sum_{\substack{\mathbf{x}\in\Lambda(\alpha_1,\dots,\alpha_{n-1},\eta) \\ \mathbf{x}\neq\mathbf{0}}} f(\mathbf{x})\,d\alpha_1 d\alpha_2 \dots d\alpha_{n-1}$$

$$= \int_{-\infty}^{\infty} \int_{-\infty}^{\infty} \dots \int_{-\infty}^{\infty} f(\mathbf{x})\,dx_1 dx_2 \dots dx_n. \tag{6}$$

Proof. It is convenient to write $f(\Lambda) = \sum_{\substack{\mathbf{x}\in\Lambda \\ \mathbf{x}\neq\mathbf{0}}} f(\mathbf{x})$.

The general point of the lattice $\Lambda(\alpha_1, \alpha_2, \ldots, \alpha_{n-1}; \eta)$ is

$$(\{u_1+u_n\alpha_1\}\chi, \{u_2+u_n\alpha_2\}\chi, \ldots, \{u_{n-1}+u_n\alpha_{n-1}\}\chi, u_n\eta),$$

where $u_1, u_2, \ldots, u_n$ are integers. If $u_n = 0$, the point reduces to

$$\mathbf{x} = (u_1\chi, u_2\chi, \ldots, u_{n-1}\chi, 0),$$

where $\chi = \eta^{-1/(n-1)}$; so that $f(\mathbf{x}) = 0$, if η is sufficiently small, and $u_1, u_2, \ldots, u_{n-1}$ are not all zero. Thus, for all sufficiently small values of η we have

$$\begin{aligned} f(\Lambda) &= f\{\Lambda(\alpha_1, \alpha_2, \ldots, \alpha_{n-1}; \eta)\} \\ &= \sum_{\substack{\mathbf{x} \in \Lambda(\alpha_1, \ldots, \alpha_{n-1}, \eta) \\ \mathbf{x} \neq 0}} f(\mathbf{x}) \\ &= \sum_{v \neq 0} \sum_{u_1=-\infty}^{\infty} \sum_{u_2=-\infty}^{\infty} \sum_{u_{n-1}=-\infty}^{\infty} f(\{u_1+v\alpha_1\}\chi, \ldots, \\ &\qquad \{u_{n-1}+v\alpha_{n-1}\}\chi, v\eta). \end{aligned}$$

Now for each v we can write

$$u_i = v\{v_i + r_i\} \quad (i = 1, 2, \ldots, n-1),$$

where v_i takes all integral values and r_i takes the $|v|$ rational values

$$0, \frac{1}{|v|}, \frac{2}{|v|}, \ldots, \frac{|v|-1}{|v|}.$$

Thus

$$\begin{aligned} f(\Lambda) = \sum_{v \neq 0} \sum_{r_1, \ldots, r_{n-1}} \sum_{v_1, \ldots, v_{n-1}} & f(\{v_1+\alpha_1+r_1\}v\chi, \ldots, \\ & \{v_{n-1}+\alpha_{n-1}+r_{n-1}\}v\chi, v\eta). \end{aligned} \tag{7}$$

Now if $v > 0$ we have

$$\begin{aligned} &\int_0^1 \int_0^1 \cdots \int_0^1 f(\{v_1+\alpha_1+r_1\}v\chi, \ldots, \{v_{n-1}+\alpha_{n-1}+r_{n-1}\}v\chi, v\eta) \\ &\qquad \times d\alpha_1\, d\alpha_2 \ldots d\alpha_{n-1} \\ &= |v\chi|^{-(n-1)} \int_{(v_1+r_1)v\chi}^{(v_1+1+r_1)v\chi} \cdots \int_{(v_{n-1}+r_{n-1})v\chi}^{(v_{n-1}+1+r_{n-1})v\chi} f(x_1, \ldots, x_{n-1}, v\eta) \\ &\qquad \times dx_1 \ldots dx_{n-1}, \end{aligned}$$

for almost all values of η; and a similar formula holds if $v < 0$. Summing these results we have

$$\int_0^1 \int_0^1 \dots \int_0^1 f\{\Lambda(\alpha_1, \dots, \alpha_{n-1}; \eta)\} \, d\alpha_1 d\alpha_2 \dots d\alpha_{n-1}$$

$$= \sum_{v \neq 0} \sum_{r_1, \dots, r_{n-1}} \sum_{v_1, \dots, v_{n-1}} \int_0^1 \int_0^1 \dots \int_0^1$$

$$\times f(\{v_1 + \alpha_1 + r_1\} v\chi, \dots, \{v_{n-1} + \alpha_{n-1} + r_{n-1}\} v\chi, v\eta)$$

$$\times \, d\alpha_1 \, d\alpha_2 \dots d\alpha_{n-1}$$

$$= \sum_{v \neq 0} \sum_{r_1, \dots, r_{n-1}} |v\chi|^{-(n-1)} \int_{-\infty}^{\infty} \dots \int_{-\infty}^{\infty} f(x_1, \dots, x_{n-1}, v\eta) \, dx_1 \dots dx_{n-1}$$

$$= \sum_{v \neq 0} \eta \int_{-\infty}^{\infty} \dots \int_{-\infty}^{\infty} f(x_1, \dots, x_{n-1}, v\eta) \, dx_1 \dots dx_{n-1},$$

for almost all η.

Hence, by the lemma,

$$\underset{\eta \to +0}{\mathrm{l.i.m.}} \int_0^1 \int_0^1 \dots \int_0^1 f\{\Lambda(\alpha_1, \alpha_2, \dots, \alpha_{n-1}; \eta)\} \, d\alpha_1 d\alpha_2 \dots d\alpha_{n-1}$$

$$= \int_{-\infty}^{\infty} \left\{ \int_{-\infty}^{\infty} \dots \int_{-\infty}^{\infty} f(x_1, \dots, x_{n-1}, x_n) \, dx_1 \dots dx_{n-1} \right\} dx_n$$

$$= \int_{-\infty}^{\infty} \int_{-\infty}^{\infty} \dots \int_{-\infty}^{\infty} f(\mathbf{x}) \, dx_1 dx_2 \dots dx_n,$$

as required.

This result shows that if η is small the average over all the lattices of the set $\Lambda(\eta)$ of the sum of $f(\mathbf{x})$, taken over all lattice points $\mathbf{x}$ other than $\mathbf{o}$, is approximately equal to the integral of $f(\mathbf{x})$ over the whole space.

2. Existence theorems for lattice packings

The main purpose of this section is to show how the following theorem follows from Theorem 4.1 of the last section.

THEOREM 4.2. *If K is a bounded set, and $\mathsf{D}K$ has a positive measure,*

$$\delta_L(K) \geqslant \frac{2\mu(K)}{\mu(\mathsf{D}K)}. \tag{8}$$

The differences between the conditions imposed on K in

this theorem and those required in Theorem 2.1 are unimportant (being chosen in each case to fit the proof adopted); the main difference is that we now prove the existence of reasonably close lattice packings, where previously we could only prove the existence of correspondingly close packings which might, however, as far as we know, be all very irregular.

Precisely as Theorem 2.2 and 2.3 follow from Theorem 2.1 in Chapter 2, so the following theorems follow from Theorem 4.2.

THEOREM 4.3. *If K is a convex body with* $\mathbf{o}$ *as centre*

$$\delta_L(K) \geqslant (\tfrac{1}{2})^{n-1}. \tag{9}$$

THEOREM 4.4. *If K is a convex body*

$$\delta_L(K) \geqslant \frac{2(n!)^2}{(2n)!}. \tag{10}$$

Proof of Theorem 4.2. Let λ be any positive real number for which

$$\lambda^n \mu(\mathsf{D}K) < 2.$$

Let $\chi(x)$ be the characteristic function of the set $\lambda \mathsf{D}K$. Then

$$\int \chi(\mathbf{x})\,d\mathbf{x} = \mu(\lambda \mathsf{D}K) = \lambda^n \mu(\mathsf{D}K) < 2.$$

So by Theorem 4.1

$$\underset{\eta\to+0}{\text{l.i.m.}} \int_0^1\int_0^1 \cdots \int_0^1 \chi\{\Lambda(\alpha_1,\ldots,\alpha_{n-1};\eta)\}\,d\alpha_1 d\alpha_2\ldots d\alpha_{n-1}$$
$$= \int \chi(\mathbf{x})\,d\mathbf{x} < 2.$$

Thus we can choose $\eta > 0$ so that

$$\frac{1}{\eta}\int_0^\eta \left[\int_0^1\int_0^1 \cdots \int_0^1 \chi\{\Lambda(\alpha_1,\ldots,\alpha_{n-1};\zeta)\}\,d\alpha_1 d\alpha_2\ldots d\alpha_{n-1}\right] d\zeta < 2.$$

Hence, we can choose ζ with $0 < \zeta < \eta$ so that

$$\int_0^1\int_0^1 \cdots \int_0^1 \chi\{\Lambda(\alpha_1,\ldots,\alpha_{n-1};\zeta)\}\,d\alpha_1 d\alpha_2\ldots d\alpha_{n-1} < 2,$$

the integral existing in the Lebesgue sense. Now we can choose $\alpha_1, \alpha_2, \ldots, \alpha_{n-1}$, satisfying

$$0 \leqslant \alpha_1 < 1, \quad 0 \leqslant \alpha_2 < 1, \ldots, 0 \leqslant \alpha_{n-1} < 1,$$

so that
$$\chi\{\Lambda(\alpha_1, \ldots, \alpha_{n-1}; \zeta)\} < 2.$$

We deduce that no point of $\Lambda = \Lambda(\alpha_1, \ldots, \alpha_{n-1}; \zeta)$ other than $\mathbf{o}$ lies in $\lambda\mathsf{D}K$. For if $\mathbf{a}$ were such a point, then by the symmetry of Λ and $\lambda\mathsf{D}K$ the point $-\mathbf{a}$ would be a second such point, and we would have

$$\chi(\Lambda) = \sum_{\substack{\mathbf{x} \in \Lambda \\ \mathbf{x} \neq \mathbf{o}}} \chi(\mathbf{x}) \geqslant \chi(\mathbf{a}) + \chi(-\mathbf{a}) = 2.$$

Consider the system $\mathscr{H}$ of translates $H + \mathbf{a}_i$ $(i = 1, 2, \ldots)$ of the set $H = \lambda K$ by the vectors $\mathbf{a}_1, \mathbf{a}_2, \ldots$ of the lattice Λ. Suppose two of these sets, say $H + \mathbf{a}_i$, $H + \mathbf{a}_j$ $(i \neq j)$, had a common point. Then there would be points $\mathbf{k}_i$ and $\mathbf{k}_j$ in K such that

$$\lambda\mathbf{k}_i + \mathbf{a}_i = \lambda\mathbf{k}_j + \mathbf{a}_j.$$

Then the point
$$\mathbf{a}_i - \mathbf{a}_j = \lambda(\mathbf{k}_j - \mathbf{k}_i)$$
would be a point of Λ, other than $\mathbf{o}$, in the set $\lambda\mathsf{D}K$. Consequently, no two of the sets can have a common point, and $\mathscr{H}$ is a lattice packing of H.

By Theorem 1.6
$$\rho_+(\mathscr{H}) = \rho_-(\mathscr{H}) = \frac{\mu(H)}{d(\Lambda)} = \mu(H). \tag{11}$$

This shows that

$$\delta_L(K) = \delta_L(\mathscr{H}) \geqslant \rho_+(\mathscr{H}) = \mu(H) = \mu(\lambda K) = \lambda^n \mu(K).$$

Since the only condition on the positive real number λ is that

$$\lambda^n \mu(\mathsf{D}K) < 2,$$

it follows that
$$\delta_L(K) \geqslant \frac{2\mu(K)}{\mu(\mathsf{D}K)}.$$

CHAPTER 5

THE EXISTENCE OF REASONABLY ECONOMICAL LATTICE COVERINGS

1. Approximation of convex sets by cylinders

In this chapter we prove the existence of some reasonably economical coverings of space, with convex bodies and with spheres. For this purpose we need a number of preliminary results, the first a result of Macbeath (1951) concerning the approximation of convex bodies by cylinders.

THEOREM 5.1. *If K is a convex body there is a cylinder H, inscribed in K, with measure satisfying*

$$\mu(H) \geqslant \frac{(n-1)^{n-1}}{n^n} \mu(K). \tag{1}$$

Proof. Let $\mathbf{a}$ and $\mathbf{b}$ be points of contact of K with a pair of parallel tac-planes. After a suitable affine transformation has been applied to K, we may suppose that the tac-planes have equations

$$x_n = 0 \quad \text{and} \quad x_n = 1,$$

and that these planes touch K at the points

$$\mathbf{o} = (0, 0, \ldots, 0) \quad \text{and} \quad \mathbf{e} = (0, 0, \ldots, 1),$$

respectively. For $0 \leqslant \lambda \leqslant 1$ let K_λ denote the section of K by the plane with equation

$$x_n = \lambda,$$

and let $\nu(K_\lambda)$ denote the $(n-1)$-dimensional Lebesgue measure (or volume) of this section. Choose ξ, with $0 \leqslant \xi \leqslant 1$, so that $\nu(K_\lambda)$ attains its maximum value for $\lambda = \xi$. Then clearly

$$\nu(K_\xi) = \int_0^1 \nu(K_\xi)\, d\lambda \geqslant \int_0^1 \nu(K_\lambda)\, d\lambda = \mu(K). \tag{2}$$

Consider the cylinder H consisting of all points of the form

$$\left(\left\{1-\frac{1}{n}\right\} x_1, \ldots, \left\{1-\frac{1}{n}\right\} x_{n-1}, \zeta\right),$$

where the $x_1, \ldots, x_{n-1}$ vary subject to

$$(x_1, \ldots, x_{n-1}, \xi) \in K_\xi,$$

while ζ varies subject to

$$\xi - \frac{1}{n}\xi \leqslant \zeta \leqslant \xi + \frac{1}{n}(1-\xi).$$

Each such point is of the form

$$\left(1-\frac{1}{n}\right)(x_1, \ldots, x_{n-1}, \xi) + \frac{1}{n}(0, \ldots, 0, n\zeta - (n-1)\xi),$$

where

$$0 \leqslant n\zeta - (n-1)\xi \leqslant 1.$$

Now

$$(x_1, \ldots, x_{n-1}, \xi) \in K \quad \text{and} \quad (0, \ldots, 0, n\zeta - (n-1)\zeta) \in K.$$

Hence, H is a cylinder contained in K with height $1/n$, and with a cross-sectional measure equal to

$$\nu\left(\left\{1-\frac{1}{n}\right\}K_\xi\right) = \left\{1-\frac{1}{n}\right\}^{n-1}\nu(K_\xi).$$

Thus, using (2),

$$\mu(H) = \frac{1}{n}\left\{1-\frac{1}{n}\right\}^{n-1}\nu(K_\xi) \geqslant \frac{(n-1)^{n-1}}{n^n}\mu(K).$$

Applying the inverse of the original affine transformation to H we obtain a cylinder (not in general a right cylinder) inscribed in the original K and satisfying the condition (1).

2. Approximation of convex sets and spheres by generalized cylinders

If S is a set in m-dimensional space, and T is a set in n-dimensional space, the Cartesian product of S and T is defined to be the set, in $(n+m)$-dimensional space, of all points

$$(x_1, x_2, \ldots, x_m, y_1, y_2, \ldots, y_n),$$

where $(x_1, x_2, \ldots, x_m) \in S$ and $(y_1, y_2, \ldots, y_n) \in T$

and is denoted by $S \times T$. This is a natural generalization of the process used in constructing a cylinder.

THEOREM 5.2. *Let K be a convex body, and let k be a positive integer less than n. After a suitable affine transformation has been applied to K, it will contain a Cartesian product $H \times C$, where H is a convex body in $(n-k)$-dimensional space, C is a cube in k-dimensional space, and*

$$\mu(H \times C) \geqslant \frac{(n-k)^{n-k}}{n^n} \mu(K). \tag{3}$$

Proof. By applying Theorem 5.1 just k times, we easily obtain such sets H, C with

$$\mu(H \times C) \geqslant \frac{(n-1)^{n-1}}{n^n} \frac{(n-2)^{n-2}}{(n-1)^{n-1}} \cdots \frac{(n-k)^{n-k}}{(n-k+1)^{n-k+1}} \mu(K)$$
$$= \frac{(n-k)^{n-k}}{n^n} \mu(K).$$

THEOREM 5.3. *Let K be a sphere in n-dimensional space, and let k be a positive integer less than n. Then K contains a Cartesian product $H \times C$, where H is an $(n-k)$-dimensional sphere, C is a k-dimensional cube, and*

$$\mu(H \times C) \geqslant \left(\frac{4}{\pi}\right)^{\frac{1}{2}k} \frac{(n-k)^{\frac{1}{2}(n-k)}}{n^{\frac{1}{2}n}} \frac{\Gamma(1+\frac{1}{2}n)}{\Gamma(1+\frac{1}{2}\{n-k\})} \mu(K). \tag{4}$$

Proof. It suffices to consider the case when K is the sphere with centre $\mathbf{o}$ and unit radius. In this case we may take $H \times C$ to be the set of points $(x_1, x_2, \ldots, x_n)$ satisfying

$$x_1^2 + \ldots + x_{n-k}^2 \leqslant 1 - \frac{k}{n},$$
$$|x_{n-k+1}| \leqslant \frac{1}{\sqrt{n}}, \ldots, |x_n| \leqslant \frac{1}{\sqrt{n}}.$$

That the result (4) holds with equality follows from the formula

$$\frac{\pi^{\frac{1}{2}h}}{\Gamma(1+\frac{1}{2}h)} \tag{5}$$

for the volume of the unit sphere in h-dimensional space.

3. Lattice covering of nearly half of space

In this section we use Theorem 4.1 to establish the existence of certain reasonably economical lattice coverings of a proportion of space not exceeding a half. We use the notation $\epsilon_+(\mathscr{K})$ of Chapter 3, § 1.

THEOREM 5.4. *Let K be a bounded set with positive measure, and let ρ, ϵ be numbers with $0 < \rho \leqslant 1$, $0 < \epsilon$. Then there is a system $\mathscr{K}$ of translates of K by the vectors of a lattice, with*

$$\rho_+(\mathscr{K}) = \rho_-(\mathscr{K}) = \rho \tag{6}$$

and

$$\epsilon_+(\mathscr{K}) < 1-\rho+\tfrac{1}{2}\rho^2+\epsilon. \tag{7}$$

Proof. Choose λ to be the positive number satisfying

$$\lambda^n\mu(K) = \rho. \tag{8}$$

Write

$$H = \lambda K,$$

so that

$$\mu(H) = \lambda^n\mu(K) = \rho.$$

We consider the system $\mathscr{H} = \mathscr{H}(\alpha_1, \ldots, \alpha_{n-1}; \eta)$ of translates of H by the vectors of the lattice $\Lambda = \Lambda(\alpha_1, \ldots, \alpha_{n-1}; \eta)$ defined in Chapter 4, § 1. Since the determinant of the lattice is 1, it follows by Theorem 1.6 that

$$\rho_+(\mathscr{H}) = \rho_-(\mathscr{H}) = \mu(H) = \rho. \tag{9}$$

Let $\chi(\mathbf{x})$ be the characteristic function of H, and let $\epsilon(\mathbf{x}) = \epsilon_\Lambda(\mathbf{x})$ be the characteristic function of the set of points belonging to no set of the system $\mathscr{H}$. We compare the function $\epsilon_\Lambda(\mathbf{x})$ with the function $\tau_\Lambda(\mathbf{x})$ defined by

$$\tau_\Lambda(\mathbf{x}) = 1-\sum_{\mathbf{a}\in\Lambda}\chi(\mathbf{x}-\mathbf{a})+\tfrac{1}{2}\sum_{\substack{\mathbf{a}\in\Lambda,\,\mathbf{b}\in\Lambda\\ \mathbf{a}\neq\mathbf{b}}}\chi(\mathbf{x}-\mathbf{a})\chi(\mathbf{x}-\mathbf{b}). \tag{10}$$

Suppose that a point $\mathbf{x}$ lies in just k (where $k \geqslant 0$) of the sets of the system H. Then $\chi(\mathbf{x}-\mathbf{a})$ takes the value 1 for just k points $\mathbf{a}$ of Λ, and takes the value 0 for the other points of Λ. Hence

$$\begin{aligned}\tau_\Lambda(\mathbf{x}) &= 1-k+\tfrac{1}{2}k(k-1)\\ &= \tfrac{1}{2}(k-1)(k-2),\end{aligned}$$

so that

$$\begin{aligned}\tau_\Lambda(\mathbf{x}) &= 1 \quad \text{if} \quad k = 0,\\ \tau_\Lambda(\mathbf{x}) &\geqslant 0 \quad \text{if} \quad k \geqslant 1.\end{aligned}$$

But

$$\begin{aligned}\epsilon_\Lambda(\mathbf{x}) &= 1 \quad \text{if} \quad k = 0,\\ \epsilon_\Lambda(\mathbf{x}) &= 0 \quad \text{if} \quad k \geqslant 1.\end{aligned}$$

Thus, in all cases $$\epsilon_\Lambda(\mathbf{x}) \leqslant \tau_\Lambda(\mathbf{x}). \tag{11}$$

Let $P = P(\Lambda)$ be the parallepiped of points of the form

$$\mathbf{x} = \nu_1 \mathbf{a}_1 + \nu_2 \mathbf{a}_2 + \ldots + \nu_n \mathbf{a}_n,$$
$$0 \leqslant \nu_1 < 1,\ 0 \leqslant \nu_2 < 1, \ldots,\ 0 \leqslant \nu_n < 1,$$

where $\mathbf{a}_1, \mathbf{a}_2, \ldots, \mathbf{a}_n$ are the points (4.1). From the periodicity of the set of points belonging to no set of $\mathscr{H}$ we have

$$\epsilon_+(\mathscr{H}) = \frac{1}{\mu(P)} \int_P \epsilon_\Lambda(\mathbf{x})\, d\mathbf{x},$$

using an abbreviated notation for integration with

$$d\mathbf{x} = dx_1 dx_2 \ldots dx_n.$$

Since $\mu(P) = 1$, it follows by (11) and (10) that

$$\begin{aligned}
\epsilon_+(\mathscr{H}) &\leqslant \int_P \tau_\Lambda(\mathbf{x})\, d\mathbf{x} \\
&= \int_P \{1 - \sum_{\mathbf{a}\in\Lambda} \chi(\mathbf{x}-\mathbf{a}) + \tfrac{1}{2} \sum_{\substack{\mathbf{a}\in\Lambda,\, \mathbf{b}\in\Lambda \\ \mathbf{a}\neq\mathbf{b}}} \chi(\mathbf{x}-\mathbf{a})\chi(\mathbf{x}-\mathbf{b})\}\, d\mathbf{x} \\
&= 1 - \sum_{\mathbf{a}\in\Lambda} \int_P \chi(\mathbf{x}-\mathbf{a})\, d\mathbf{x} \\
&\quad + \tfrac{1}{2} \sum_{\mathbf{a}\in\Lambda} \int_P \sum_{\substack{\mathbf{b}\in\Lambda \\ \mathbf{b}\neq\mathbf{a}}} \chi(\mathbf{x}-\mathbf{a})\chi(\mathbf{x}-\mathbf{b})\, d\mathbf{x}.
\end{aligned}$$

Writing $\mathbf{x} = \mathbf{a}+\mathbf{y}$, $\mathbf{b} = \mathbf{a}+\mathbf{c}$, and using the fact that the sets

$$P - \mathbf{a} \quad (\mathbf{a} \in \Lambda)$$

fit together to make up the whole space, we obtain

$$\begin{aligned}
\epsilon_+(\mathscr{H}) &\leqslant 1 - \sum_{\mathbf{a}\in\Lambda} \int_{P-\mathbf{a}} \chi(\mathbf{y})\, d\mathbf{y} \\
&\quad + \tfrac{1}{2} \sum_{\mathbf{a}\in\Lambda} \int_{P-\mathbf{a}} \sum_{\substack{\mathbf{c}\in\Lambda \\ \mathbf{c}\neq\mathbf{o}}} \chi(\mathbf{y})\chi(\mathbf{y}-\mathbf{c})\, d\mathbf{y} \\
&= 1 - \int \chi(\mathbf{y})\, d\mathbf{y} + \tfrac{1}{2} \int \sum_{\substack{\mathbf{c}\in\Lambda \\ \mathbf{c}\neq\mathbf{o}}} \chi(\mathbf{y})\chi(\mathbf{y}-\mathbf{c})\, d\mathbf{y} \\
&= 1 - \rho + \tfrac{1}{2} \sum_{\substack{\mathbf{c}\in\Lambda \\ \mathbf{c}\neq\mathbf{o}}} \int \chi(\mathbf{y})\chi(\mathbf{y}-\mathbf{c})\, d\mathbf{y}.
\end{aligned}$$

Now the function

$$\int \chi(\mathbf{y})\chi(\mathbf{y}-\mathbf{z})d\mathbf{y}$$

is integrable in the Lebesgue sense with respect to $\mathbf{z}$ over the whole of space, and vanishes outside a bounded region. So, by Theorem 4.1, we have

$$\begin{aligned}
\underset{\eta\to+0}{\text{l.i.m.}} \int_0^1 \dots \int_0^1 \epsilon_+(\mathscr{H})\,d\alpha_1 \dots d\alpha_{n-1} \\
&= 1-\rho+\frac{1}{2}\int\left[\int \chi(\mathbf{y})\chi(\mathbf{y}-\mathbf{z})d\mathbf{y}\right]d\mathbf{z} \\
&= 1-\rho+\frac{1}{2}\int\left[\int \chi(\mathbf{y})\chi(\mathbf{y}-\mathbf{z})d\mathbf{z}\right]d\mathbf{y} \\
&= 1-\rho+\frac{1}{2}\int\left[\chi(\mathbf{y})\int \chi(-\mathbf{x})d\mathbf{x}\right]d\mathbf{y} \\
&= 1-\rho+\tfrac{1}{2}\rho^2,
\end{aligned}$$

on changing the order of integration, and writing $\mathbf{z} = \mathbf{x}+\mathbf{y}$. Now, by appropriate choice of η and $\alpha_1, \dots, \alpha_{n-1}$, we can ensure that

$$\epsilon_+(\mathscr{H}) = \epsilon_+\{\mathscr{H}(\alpha_1, \dots, \alpha_{n-1}; \eta)\} < 1-\rho+\tfrac{1}{2}\rho^2+\epsilon.$$

Finally, we take $\mathscr{K}$ to be the system of translates of $K = \lambda^{-1}H$ by the vectors of

$$\lambda^{-1}\Lambda = \lambda^{-1}\Lambda(\alpha_1, \dots, \alpha_{n-1}; \eta).$$

From considerations of homogeneity

$$\rho_+(\mathscr{K}) = \rho_-(\mathscr{K}) = \rho_+(\mathscr{H}) = \rho_-(\mathscr{H}) = \rho$$

and

$$\epsilon_+(\mathscr{K}) = \epsilon_+(\mathscr{H}) < 1-\rho+\tfrac{1}{2}\rho^2+\epsilon$$

as required.

4. Lattice covering of most of space by generalized cylinders

In this section we prove a lemma which enables us to obtain, from a lattice covering of a proportion of n-dimensional space by the translates of a set H, a rather less efficient lattice covering of a higher proportion of $(n+1)$-dimensional space by cylinders with H as base. By repeated applications of the lemma, we will be able to obtain lattice coverings of most of space by generalized cylinders.

THEOREM 5.5. *Let H be a set in n-dimensional space which is bounded and which has positive measure. Let $\mathscr{H}$ be the system of translates of H by the vectors of a lattice Λ. Let C be the 1-dimensional segment of height 2, and let $K = H \times C$. Then there is an $(n+1)$-dimensional lattice Λ_1, such that the system $\mathscr{K}$ of translates of K by the vectors of Λ_1 satisfies*

$$\rho_+(\mathscr{K}) = \rho_-(\mathscr{K}) = 2\rho_+(\mathscr{H}) = 2\rho_-(\mathscr{H}), \tag{12}$$

$$\epsilon_+(\mathscr{K}) \leqslant [\epsilon_+(\mathscr{H})]^2. \tag{13}$$

Proof. The proof is based on the general idea used in the proof of Theorem 3.1. It is convenient to assume that Λ is the lattice of points with integral coordinates; the general case is easily reduced to this special case by application of the appropriate linear transformation to H and Λ.

Now let $\epsilon(\mathbf{x})$ be the characteristic function of the set of points which belong to no set of the system $\mathscr{H}$. Then $\epsilon(\mathbf{x})$ is periodic in each coordinate with unit period. Hence

$$\epsilon_+(\mathscr{H}) = \int_0^1 \dots \int_0^1 \epsilon(\mathbf{x})\, dx_1 \dots dx_n.$$

So, writing $\mathbf{z} = \mathbf{x} - \mathbf{y}$, and using the periodicity, we have

$$\int_0^1 \dots \int_0^1 \left\{\int_0^1 \dots \int_0^1 \epsilon(\mathbf{x})\epsilon(\mathbf{x}-\mathbf{y})\, dx_1 \dots dx_n\right\} dy_1 \dots dy_n$$

$$= \int_0^1 \dots \int_0^1 \epsilon(\mathbf{x}) \left\{\int_0^1 \dots \int_0^1 \epsilon(\mathbf{x}-\mathbf{y})\, dy_1 \dots dy_n\right\} dx_1 \dots dx_n$$

$$= \int_0^1 \dots \int_0^1 \epsilon(\mathbf{x}) \left\{\int_{x_1-1}^{x_1} \dots \int_{x_n-1}^{x_n} \epsilon(\mathbf{z})\, dz_1 \dots dz_n\right\} dx_1 \dots dx_n$$

$$= \int_0^1 \dots \int_0^1 \epsilon(\mathbf{x}) \left\{\int_0^1 \dots \int_0^1 \epsilon(\mathbf{z})\, dz_1 \dots dz_n\right\} dx_1 \dots dx_n$$

$$= \int_0^1 \dots \int_0^1 \epsilon(\mathbf{x})\, \epsilon_+(\mathscr{H})\, dx_1 \dots dx_n = [\epsilon_+(\mathscr{H})]^2.$$

Hence, we can choose a point **a** such that

$$\int_0^1 \dots \int_0^1 \epsilon(\mathbf{x})\,\epsilon(\mathbf{x}-\mathbf{a})\,dx_1 \dots dx_n \leqslant [\epsilon_+(\mathscr{H})]^2. \tag{14}$$

Take $K = H \times C$, where C is the segment

$$0 \leqslant x_{n+1} < 2$$

of height 2. Take Λ_1 to be the lattice generated by the points

$$\begin{aligned} \mathbf{e}_1 &= (1, 0, \dots, 0, 0), \\ \mathbf{e}_2 &= (0, 1, \dots, 0, 0), \\ &\dots\dots\dots\dots\dots\dots \\ \mathbf{e}_n &= (0, 0, \dots, 1, 0), \\ \mathbf{a} + \mathbf{e}_{n+1} &= (a_1, a_2, \dots, a_n, 1). \end{aligned}$$

Take $\mathscr{K}$ to be the system of translates of K by the vectors of Λ_1. For each integer u, the part of the system $\mathscr{K}$ lying between the planes

$$x_{n+1} = u, \quad x_{n+1} = u+1,$$

differs from that lying between the planes

$$x_{n+1} = 0, \quad x_{n+1} = 1,$$

solely by translation by the vector

$$u(a_1, a_2, \dots, a_n, 1).$$

So it suffices to study that part of the system $\mathscr{K}$ lying in the strip defined by

$$1 \leqslant x_{n+1} < 2.$$

The sets of $\mathscr{K}$ which intersect this strip are solely the sets

$$K + (u_1, \dots, u_n, 0),$$

where $u_1, \dots, u_n$ are integers, and the sets

$$K + (u_1 + a_1, \dots, u_n + a_n, 1),$$

where again $u_1, \dots, u_n$ are integers. The intersections of these sets with the strip are

$$(H + \mathbf{u}) \times C' \quad (\mathbf{u} \in \Lambda),$$

and

$$(H + \mathbf{u} + \mathbf{a}) \times C' \quad (\mathbf{u} \in \Lambda),$$

where C' is the segment
$$1 \leqslant x_{n+1} < 2.$$
If $\chi(x_{n+1})$ is the characteristic function of C', the characteristic function of the set of points that are in the strip, but do not belong to any of the sets
$$(H+\mathbf{u}) \times C' \quad (\mathbf{u} \in \Lambda),$$
is
$$\epsilon(\mathbf{x})\,\chi(x_{n+1}),$$
while the characteristic function of the set of points that are in the strip but do not belong to any of the sets
$$(H+\mathbf{u}+\mathbf{a}) \times C' \quad (\mathbf{u} \in \Lambda),$$
is
$$\epsilon(\mathbf{x}-\mathbf{a})\,\chi(x_{n+1}).$$
Thus the characteristic function of the set of points that belong to the strip but do not belong to any set of the system $\mathscr{K}$, is
$$\epsilon(\mathbf{x})\,\epsilon(\mathbf{x}-\mathbf{a})\,\chi(x_{n+1}).$$

From the periodicity of the system $\mathscr{K}$ we deduce that
$$\begin{aligned}\epsilon_+(\mathscr{K}) &= \int_1^2 \left\{\int_0^1 \cdots \int_0^1 \epsilon(\mathbf{x})\,\epsilon(\mathbf{x}-\mathbf{a})\,\chi(x_{n+1})\,dx_1 \ldots dx_n\right\} dx_{n+1} \\ &= \int_0^1 \int_0^1 \cdots \int_0^1 \epsilon(\mathbf{x})\,\epsilon(\mathbf{x}-\mathbf{a})\,dx_1\,dx_2 \ldots dx_n \\ &\leqslant [\epsilon_+(\mathscr{H})]^2,\end{aligned}$$
by (14).

By Theorem 1.6 we have
$$\begin{aligned}\rho_+(\mathscr{K}) = \rho_-(\mathscr{K}) = \mu(K) &= \mu(H \times C) \\ &= 2\mu(H) = 2\rho_+(\mathscr{H}) = 2\rho_-(\mathscr{H})\end{aligned}$$
as required.

THEOREM 5.6. *Let H be a bounded set with positive measure in $(n-k)$-dimensional space, where $1 \leqslant k < n-1$. Let C_k be a k-dimensional cube, and let $K = H \times C_k$. Then, for each $\eta > 0$, there is a lattice Λ, such that the system $\mathscr{K}$ of translates of K by the vectors of Λ satisfies*
$$\rho_+(\mathscr{K}) = \rho_-(\mathscr{K}) = 2^k, \tag{15}$$
$$\epsilon_+(\mathscr{K}) < (\tfrac{1}{2})^{2^k} + \eta. \tag{16}$$

Proof. By Theorem 5.4, with $\rho = 1$, for each $\epsilon > 0$ there is a lattice Λ_0, such that the system $\mathscr{H} = \mathscr{K}_0$ of translates of $H = K_0$ by the vectors of Λ_0 satisfies

$$\rho_+(\mathscr{K}_0) = \rho_-(\mathscr{K}_0) = 1,$$
$$\epsilon_+(\mathscr{K}_0) = \tfrac{1}{2} + \epsilon.$$

Let C_r ($r = 1, 2, \ldots, k$) denote the r-dimensional cube

$$0 \leqslant x_{n-k+1} < 1,$$
$$\cdots\cdots\cdots\cdots\cdots$$
$$0 \leqslant x_{n-k+r} < 1,$$

and write
$$K_r = K_0 \times C_r.$$

Then, using Theorem 5.5 inductively, there will be lattices $\Lambda_1, \Lambda_2, \ldots, \Lambda_k$, such that the corresponding systems

$$\mathscr{K}_1, \mathscr{K}_2, \ldots, \mathscr{K}_k$$

of translates of K_1, $K_2, \ldots, K_k$ by the vectors of $\Lambda_1, \Lambda_2, \ldots, \Lambda_k$ satisfy

$$\rho_+(\mathscr{K}_r) = \rho_-(\mathscr{K}_r) = 2^r,$$
$$\epsilon_+(\mathscr{K}_r) = (\tfrac{1}{2} + \epsilon)^{2^r}$$

for $r = 1, 2, \ldots, k$. The result now follows on taking $\Lambda = \Lambda_k$, $\mathscr{K} = \mathscr{K}_k$ and ϵ sufficiently small.

5. Lattice covering of space with convex sets and spheres

Before we prove our main theorems we need one more preliminary result showing that we can obtain a lattice covering of the whole of space, once we have a lattice covering of a sufficiently large proportion of space.

THEOREM 5.7. *Let $\mathscr{K}$ be the system of translates of a convex body K by the vectors of a lattice Λ, and suppose that*

$$\epsilon_+(\mathscr{K}) < \frac{1}{h^n + 1}, \tag{17}$$

where h is a positive integer. Then the system $\mathscr{H}$ of translates of the body $H = (1 + h^{-1}) K$ by the vectors of Λ is a lattice covering of space with

$$\rho_+(\mathscr{H}) = \rho_-(\mathscr{H}) = (1 + h^{-1})^n \rho_+(\mathscr{K}) = (1 + h^{-1})^n \rho_-(\mathscr{K}). \tag{18}$$

Proof. It is convenient to take Λ to be the lattice of points with integral coordinates. We regard Λ as an additive group, and split it into its h^n cosets modulo the subgroup $h\Lambda$. Let these be

$$\Lambda_1 = h\Lambda, \Lambda_2, \ldots, \Lambda_{h^n}.$$

Then Λ_1 is the lattice of all points whose coordinates are integral multiples of h, while for each i, with $1 \leqslant i \leqslant h^n$, the set Λ_i is the translate of Λ_1 by a vector with integral coordinates. Take C_h to be the cube

$$0 \leqslant x_1 < h, \ldots, 0 \leqslant x_n < h,$$

and let S_i be defined, for $1 \leqslant i \leqslant h^n$, to be the intersection of C_h with the union

$$\bigcup_{\mathbf{u} \in \Lambda_i} \{K + \mathbf{u}\}.$$

Since the sets $K + \mathbf{u}$, with $\mathbf{u} \in \Lambda_i$, exhaust the system $\mathscr{K}$, when i takes the values 1, 2, ..., h^n, it follows from the periodicity of the system that

$$\sum_{i=1}^{h^n} \mu(S_i) \geqslant \mu(C_h)\{1 - \epsilon_+(\mathscr{K})\} = h^n\{1 - \epsilon_+(\mathscr{K})\}.$$

So, for some j with $1 \leqslant j \leqslant h^n$, we have

$$\mu(S_j) \geqslant 1 - \epsilon_+(\mathscr{K}) > 1 - \frac{1}{h^n + 1} = \frac{h^n}{h^n + 1}.$$

But the different sets

$$\bigcup_{\mathbf{u} \in \Lambda_i} \{K + \mathbf{u}\}$$

differ solely by a translation by a vector of Λ, and each is periodic with period h in each coordinate. Hence all the sets S_i have the same measure, and, in particular,

$$\mu(S_1) = \mu(S_j) > \frac{h^n}{h^n + 1}.$$

Now consider the system of sets

$$-h^{-1}K + \mathbf{u}$$

with $\mathbf{u}$ in Λ. This system is periodic in each coordinate with unit period. The intersection of the union

$$\bigcup_{\mathbf{u} \in \Lambda} \{-h^{-1}K + \mathbf{u}\}$$

with the cube

$$-1 < x_1 \leqslant 0, \ldots, -1 < x_n \leqslant 0,$$

is precisely the set

$$-h^{-1}S_1.$$

So, for any point $\mathbf{x}$, the measure of the intersection of the cube C_1 given by

$$0 \leqslant x_1 < 1, \ldots, 0 \leqslant x_n < 1,$$

with the union

$$\bigcup_{\mathbf{u}\in\Lambda} \{-h^{-1}K+\mathbf{x}+\mathbf{u}\}$$

is

$$\mu(h^{-1}S_1) = h^{-n}\mu(S_1) > \frac{1}{h^n+1}.$$

But the measure of the intersection of the cube C_1 with the union

$$\bigcup_{\mathbf{u}\in\Lambda} \{K+\mathbf{u}\}$$

is

$$1-\epsilon_+(\mathscr{K}) > 1-\frac{1}{h^n+1}.$$

We now have two sets, with total measure greater than 1, in the cube C_1 of measure 1. Hence the two sets have a common point. Thus there are points $\mathbf{k}_1$ and $\mathbf{k}_2$ in K, and points $\mathbf{u}_1$ and $\mathbf{u}_2$ in Λ with

$$-h^{-1}\mathbf{k}_1+\mathbf{x}+\mathbf{u}_1 = \mathbf{k}_2+\mathbf{u}_2.$$

So

$$\mathbf{x} = \mathbf{k}_2+h^{-1}\mathbf{k}_1+\mathbf{u}_2-\mathbf{u}_1,$$

and $\mathbf{x}$ belongs to the system $\mathscr{H}$ of translates of $H = (1+h^{-1})K$ by the vectors of Λ. Since this holds for all $\mathbf{x}$, the system $\mathscr{H}$ is a lattice covering of space. The result (18) follows by Theorem 1.6.

We now come to the main theorems of this chapter.

THEOREM 5.8. *There is a constant c, such that*

$$\vartheta_L(K) \leqslant n^{\log_2 n + c\log\log n}, \tag{19}$$

for each convex body K in n-dimensional space with $n \geqslant 3$.

THEOREM 5.9. *There is a constant c, such that*

$$\vartheta_L(S) \leqslant c(n\log_2 n)^{\log_2 \sqrt{(2\pi e)}}, \tag{20}$$

for each sphere S in n-dimensional space with $n \geqslant 3$.

Proof of Theorem 5.8. We take k to be the least integer satisfying

$$k > \log_2 n + \log_2\log_2 n + 1. \tag{21}$$

Then $k < n$ provided $n \geqslant 8$. By Theorem 5.2, after application of a suitable affine transformation, K will contain a Cartesian product $H \times C$, where H is a convex body in $(n-k)$-dimensional space, and C is a cube in k-dimensional space, and where

$$\mu(H \times C) \geqslant \frac{(n-k)^{n-k}}{n^n} \mu(K). \tag{22}$$

By Theorem 5.6, for each $\eta > 0$, there is a lattice Λ', such that the system $\mathscr{K}'$ of translates of $K' = H \times C$ by the vectors of Λ' satisfies

$$\rho_+(\mathscr{K}') = \rho_-(\mathscr{K}') = 2^k \tag{23}$$

and

$$\epsilon_+(\mathscr{K}') < (\tfrac{1}{2})^{2^k} + \eta.$$

Thus, provided η is sufficiently small, it follows from (21) that

$$\epsilon_+(\mathscr{K}') < (\tfrac{1}{2})^{2n \log_2 n} = \frac{1}{n^{2n}} < \frac{1}{n^n + 1}.$$

Since $H \times C$ is contained in K, it follows from Theorem 5.7, with $h = n$, that the system $\mathscr{J}$ of translates of the set

$$J = (1 + n^{-1}) K$$

by the vectors of Λ is a lattice covering of space with

$$\rho_+(\mathscr{J}) = \rho_-(\mathscr{J}) = (1 + n^{-1})^n \rho_+(\mathscr{K}'') = (1 + n^{-1})^n \rho_-(\mathscr{K}''),$$

where $\mathscr{K}''$ is the system of translates of K by the vectors of Λ. But, by Theorem 1.6, we have

$$\rho_+(\mathscr{K}'') = \rho_-(\mathscr{K}'') = \frac{\mu(K)}{\mu(K')} \rho_+(\mathscr{K}') = \frac{\mu(K)}{\mu(K')} \rho_-(\mathscr{K}').$$

Thus $\mathscr{J}$ is a lattice covering of space with

$$\begin{aligned}
\rho_+(\mathscr{J}) = \rho_-(\mathscr{J}) &= (1 + n^{-1}) \frac{\mu(K)}{\mu(H \times C)} \rho_+(\mathscr{K}') \\
&= (1 + n^{-1})^n \frac{\mu(K)}{\mu(H \times C)} 2^k \\
&\leqslant (1 + n^{-1})^n \frac{n^n}{(n-k)^{n-k}} 2^k \\
&= (1 + n^{-1})^n \left(1 + \frac{k}{n-k}\right)^{n-k} n^k 2^k \\
&\leqslant e^{k+1} n^k 2^k = n^{k + (k/\log_2 n) + \{(k+1)/\log_e n\}} \\
&\leqslant n^{\log_2 n + c \log_2 \log_2 n}
\end{aligned}$$

for $n \geqslant 8$ and c a suitable constant.

This implies that

$$\vartheta_L(K) = \vartheta_L(J) \leqslant \rho_+(\mathscr{J}) \leqslant n^{\log_2 n + c \log_2 \log_2 n}$$

for $n \geqslant 8$. If $3 \leqslant n \leqslant 7$, we easily obtain a bound for $\vartheta_L(K)$ by the above method, taking $k = 1$ and $h = 1$. So, increasing the value of c, if necessary, we see that (19) holds for $n \geqslant 3$.

Proof of Theorem 5.9. By use of the method used to prove Theorem 5.8, but using Theorem 5.3 in place of Theorem 5.2, it is easy to show that

$$\vartheta_L(S) \leqslant (1+n^{-1})^n \left(\frac{\pi}{4}\right)^{\frac{1}{2}k} \frac{n^{\frac{1}{2}n}}{(n-k)^{\frac{1}{2}(n-k)}} \frac{\Gamma(1+\frac{1}{2}(n-k))}{\Gamma(1+\frac{1}{2}n)} 2^k,$$

where $$k \leqslant \log_2 n + \log_2 \log_2 n + 2,$$

for $n \geqslant 8$. Substituting for k, and using the Stirling expansion for the gamma function, we obtain

$$\vartheta_L(S) \leqslant c(n \log_2 n)^{\log_2 \sqrt{(2\pi e)}},$$

for a suitable constant c and $n \geqslant 8$. The inequality extends to $n \geqslant 3$, on replacing c by a larger number if necessary.

CHAPTER 6

PACKINGS OF SIMPLICES CANNOT BE VERY DENSE

1. Packing of sets and their difference sets

Before we study the packing of simplices, we first establish some elementary results, due essentially to Minkowski (1904), connecting the packing densities of a set with those of its difference body.

THEOREM 6.1. *Let K be a bounded set with positive measure. Then*

$$\frac{\delta(\mathsf{D}K)}{\mu(\mathsf{D}K)} \leqslant \frac{\delta(K)}{2^n\mu(K)}, \tag{1}$$

$$\frac{\delta_L(\mathsf{D}K)}{\mu(\mathsf{D}K)} \leqslant \frac{\delta_L(K)}{2^n\mu(K)}. \tag{2}$$

Further, equality holds in (1) *and* (2), *whenever K is convex.*

Proof. Write $H = \frac{1}{2}\mathsf{D}K$, and let $\mathscr{H}$ be any system of translates

$$H+\mathbf{a}_i \quad (i = 1, 2, \ldots)$$

of H which forms a packing. Consider the corresponding system $\mathscr{K}$ of translates

$$K+\mathbf{a}_i \quad (i = 1, 2, \ldots)$$

of K. If two of the sets, say $K+\mathbf{a}_1$, $K+\mathbf{a}_2$, had a point in common, we would have

$$\mathbf{k}_1+\mathbf{a}_1 = \mathbf{k}_2+\mathbf{a}_2$$

for some points $\mathbf{k}_1$, $\mathbf{k}_2$ in K. Then the points

$$\mathbf{h}_1 = \tfrac{1}{2}(\mathbf{k}_1-\mathbf{k}_2), \quad \mathbf{h}_2 = \tfrac{1}{2}(\mathbf{k}_2-\mathbf{k}_1)$$

would belong to $H = \frac{1}{2}\mathsf{D}K$, and the point

$$\begin{aligned}\mathbf{h}_1+\mathbf{a}_1 &= \mathbf{k}_1+ \mathbf{a}_1-\tfrac{1}{2}(\mathbf{k}_1+\mathbf{k}_2)\\ &= \mathbf{k}_2+\mathbf{a}_2-\tfrac{1}{2}(\mathbf{k}_1+\mathbf{k}_2)\\ &= \mathbf{h}_2+\mathbf{a}_2\end{aligned}$$

would be common to $H+\mathbf{a}_1$ and $H+\mathbf{a}_2$, contrary to the choice of $\mathscr{H}$. Hence the system $\mathscr{K}$ is a packing. By the methods used in Chapter 1 it follows that

$$\frac{\rho_-(\mathscr{K})}{\mu(K)} = \frac{\rho_-(\mathscr{H})}{\mu(H)},$$

and

$$\frac{\rho_+(\mathscr{K})}{\mu(K)} = \frac{\rho_+(\mathscr{H})}{\mu(H)}.$$

Since we have a packing $\mathscr{K}$ of K corresponding to each packing $\mathscr{H}$ of H and a lattice packing $\mathscr{K}_L$ of K corresponding to each lattice packing $\mathscr{H}_L$ of H, it follows that

$$\frac{\delta(K)}{\mu(K)} \geqslant \frac{\delta(H)}{\mu(H)},$$

$$\frac{\delta_L(K)}{\mu(K)} \geqslant \frac{\delta_L(H)}{\mu(H)}.$$

The results (1) and (2) follow on noting that

$$\delta(H) = \delta(\tfrac{1}{2}\mathsf{D}K) = \delta(\mathsf{D}K),$$

$$\delta_L(H) = \delta_L(\tfrac{1}{2}\mathsf{D}K) = \delta_L(\mathsf{D}K),$$

$$\mu(H) = \mu(\tfrac{1}{2}\mathsf{D}K) = (\tfrac{1}{2})^n \mu(\mathsf{D}K).$$

Now if K is convex and $\mathscr{K}$ is a system of translates

$$K+\mathbf{a}_i \quad (i = 1, 2, \ldots)$$

of K, which forms a packing, we consider the corresponding system $\mathscr{H}$ of translates

$$H+\mathbf{a}_i \quad (i = 1, 2, \ldots)$$

of H. If two of the sets, say $H+\mathbf{a}_1$, $H+\mathbf{a}_2$, had a point in common we would have

$$\mathbf{h}_1+\mathbf{a}_1 = \mathbf{h}_2+\mathbf{a}_2,$$

for some points $\mathbf{h}_1$, $\mathbf{h}_2$ in H. Then we could represent $\mathbf{h}_1$ and $\mathbf{h}_2$ in the forms

$$\mathbf{h}_1 = \tfrac{1}{2}(\mathbf{k}_{11}-\mathbf{k}_{12}), \quad \mathbf{h}_2 = \tfrac{1}{2}(\mathbf{k}_{21}-\mathbf{k}_{22}),$$

where $\mathbf{k}_{11}$, $\mathbf{k}_{12}$, $\mathbf{k}_{21}$, $\mathbf{k}_{22}$ belong to K. As K is convex, the points

$$\mathbf{k}_1 = \tfrac{1}{2}(\mathbf{k}_{11}+\mathbf{k}_{22}), \quad \mathbf{k}_2 = \tfrac{1}{2}(\mathbf{k}_{12}+\mathbf{k}_{21})$$

would belong to K, and the point

$$\mathbf{k}_1+\mathbf{a}_1 = \mathbf{k}_2+\mathbf{a}_2$$

would be common to the sets $K+\mathbf{a}_1, K+\mathbf{a}_2$ contrary to the choice of $\mathscr{K}$. Thus $\mathscr{H}$ is a packing. As in the last paragraph we obtain the reversed inequalities

$$\frac{\delta(H)}{\mu(H)} \geqslant \frac{\delta(K)}{\mu(K)},$$

$$\frac{\delta_L(H)}{\mu(H)} \geqslant \frac{\delta_L(K)}{\mu(K)}.$$

Consequently, (1) and (2) hold with equality.

2. Packing of simplices

We first calculate the volume of the difference set of a simplex.

THEOREM 6.2. *If S is an n-dimensional simplex*

$$\mu(\mathsf{D}S) = \binom{2n}{n}\mu(S). \tag{3}$$

Proof. Since the ratio $\mu(\mathsf{D}S)/\mu(S)$ is independent of the choice of the simplex S, it suffices to consider the case when S is the set of points $\mathbf{x}$ satisfying

$$x_1+x_2+\ldots+x_n \leqslant 1,$$

$$x_1 \geqslant 0, \quad x_2 \geqslant 0, \ldots, x_n \geqslant 0.$$

Then
$$\mu(S) = \frac{1}{n!}.$$

Now consider the set of all points of $\mathsf{D}S$ for which r of the coordinates, say $x_{i(1)}, \ldots, x_{i(r)}$, are non-negative, while the others, say $x_{j(1)}, \ldots, x_{j(n-r)}$, are non-positive. Each such point, being a difference of points of S, clearly satisfies

$$\left.\begin{aligned} x_{i(1)}+x_{i(2)}+\ldots+x_{i(r)} &\leqslant 1, \\ x_{i(1)} \geqslant 0, x_{i(2)} \geqslant 0, \ldots, x_{i(r)} &\geqslant 0, \\ -x_{j(1)}-x_{j(2)}-\ldots-x_{j(n-r)} &\leqslant 1, \\ -x_{j(1)} \geqslant 0, -x_{j(2)} \geqslant 0, \ldots, -x_{j(n-r)} &\geqslant 0. \end{aligned}\right\} \tag{4}$$

But, on the other hand, each point satisfying the inequalities (4) is expressible as the difference between points of S; using a clumsy but explicit notation, the representation is in the form

$$\mathbf{x} = (\tfrac{1}{2}|x_1| + \tfrac{1}{2}x_1, \tfrac{1}{2}|x_2| + \tfrac{1}{2}x_2, \ldots, \tfrac{1}{2}|x_n| + \tfrac{1}{2}x_n) - (\tfrac{1}{2}|x_1| - \tfrac{1}{2}x_1, \tfrac{1}{2}|x_2| - \tfrac{1}{2}x_2, \ldots, \tfrac{1}{2}|x_n| - \tfrac{1}{2}x_n).$$

The set of points satisfying (4) is clearly a Cartesian product of two simplices, one of dimension r, the other of dimension $n-r$; its volume is

$$\frac{1}{r!}\frac{1}{(n-r)!}.$$

For fixed r the set $\mathsf{D}S$ contains just $\binom{n}{r}$ such sets of total volume

$$\binom{n}{r}\binom{n}{r}\frac{1}{n!}.$$

Hence
$$\mu(\mathsf{D}S) = \sum_{r=0}^{n}\left\{\binom{n}{r}\right\}^2\frac{1}{n!} = \binom{2n}{n}\mu(S),$$

since
$$\sum_{r=0}^{n}\left\{\binom{n}{r}\right\}^2 = \binom{2n}{n},$$

as may be seen by comparing the coefficients of x^n in the formula

$$\left(\sum_{r=0}^{n}\binom{n}{r}x^r\right)\times\left(\sum_{r=0}^{n}\binom{n}{r}x^{n-r}\right) = (1+x)^{2n}.$$

We now obtain

THEOREM 6.3. *If S is a simplex in n-dimensional space*

$$\frac{2(n!)^2}{(2n)!} \leqslant \delta_L(S) \leqslant \delta(S) \leqslant \frac{2^n(n!)^2}{(2n)!}.$$

Proof. Combining Theorems 6.1, 6.2 and 1.3 we have

$$\delta(S) = \frac{2^n\mu(S)}{\mu(\mathsf{D}S)}\delta(\mathsf{D}S) = \frac{2^n(n!)^2}{(2n)!}\delta(\mathsf{D}S) \leqslant \frac{2^n(n!)^2}{(2n)!}.$$

The inequalities

$$\frac{2(n!)^2}{(2n)!} \leqslant \delta_L(S) \leqslant \delta(S)$$

follow as particular cases of Theorems 4.4 and 1.4.

It is of interest to note that

$$\frac{2(n!)^2}{(2n)!} \sim \frac{2\sqrt{(\pi n)}}{4^n},$$

$$\frac{2^n(n!)^2}{(2n)!} \sim \frac{\sqrt{(\pi n)}}{2^n}$$

as $n \to \infty$.

CHAPTER 7

PACKINGS OF SPHERES CANNOT BE VERY DENSE

1. The Voronoi polyhedra

In this section we describe the construction of the Voronoi polyhedra associated with a distribution of points throughout space (see Voronoi, 1908). *We suppose throughout that the set of points*

$$\mathbf{a}_1, \mathbf{a}_2, \ldots$$

is discrete, and that there is a positive number R, with the property that for each point $\mathbf{x}$ of space there is a point $\mathbf{a}_i$ of the sequence, whose distance $|\mathbf{x}-\mathbf{a}_i|$ from $\mathbf{x}$ is less than R.

With each point $\mathbf{a}_i$ we associate the set $\Pi(\mathbf{a}_i)$ of all points $\mathbf{x}$ whose distance from $\mathbf{a}_i$ is equal to their minimum distance from the points of $\{\mathbf{a}_j\}$. Then $\Pi(\mathbf{a}_i)$ is the set of all points $\mathbf{x}$ satisfying

$$|\mathbf{x}-\mathbf{a}_i| \leqslant |\mathbf{x}-\mathbf{a}_j| \quad (j \neq i).$$

Using the usual scalar product notation this condition is equivalent to the condition

$$(\mathbf{x}-\mathbf{a}_i).(\mathbf{x}-\mathbf{a}_i) \leqslant (\mathbf{x}-\mathbf{a}_j).(\mathbf{x}-\mathbf{a}_j) \quad (j \neq i),$$

and so to the condition

$$\mathbf{x}.(\mathbf{a}_j-\mathbf{a}_i) \leqslant (\tfrac{1}{2}\mathbf{a}_i+\tfrac{1}{2}\mathbf{a}_j).(\mathbf{a}_j-\mathbf{a}_i) \quad (j \neq i). \tag{1}$$

Thus $\Pi(\mathbf{a}_i)$ is the intersection of all the half-spaces (1) for $j \neq i$. Geometrically (1) represents the half-space containing $\mathbf{a}_i$ bounded by the plane perpendicular to the segment $\mathbf{a}_i\mathbf{a}_j$, and passing through the mid-point $\frac{1}{2}\mathbf{a}_i+\frac{1}{2}\mathbf{a}_j$ of this segment.

Let J_i denote the set of integers j, other than i, for which

$$|\mathbf{a}_j-\mathbf{a}_i| < 2R. \tag{2}$$

Let $\Pi'(\mathbf{a}_i)$ denote the intersection of all the closed half-spaces

$$\mathbf{x}.(\mathbf{a}_j-\mathbf{a}_i) \leqslant (\tfrac{1}{2}\mathbf{a}_i+\tfrac{1}{2}\mathbf{a}_j).(\mathbf{a}_j-\mathbf{a}_i) \tag{3}$$

with j in J_i. Then $\Pi(\mathbf{a}_i)$ is clearly contained in $\Pi'(\mathbf{a}_i)$. But suppose that $\mathbf{p}$ is any point of $\Pi'(\mathbf{a}_i)$, other than $\mathbf{a}_i$ itself. Let $\mathbf{q}$ be a point on the segment $\mathbf{a}_i\mathbf{p}$, or on this segment produced beyond $\mathbf{p}$, chosen so that

$$|\mathbf{q}-\mathbf{a}_i| = R.$$

Then $\mathbf{p} = \mathbf{a}_i+\lambda(\mathbf{q}-\mathbf{a}_i)$ for some positive λ. By the supposition italicized above, there will be an integer k with

$$|\mathbf{q}-\mathbf{a}_k| < R.$$

We have
$$0 < |\mathbf{a}_i-\mathbf{a}_k| < 2R,$$
so that k belongs to J_i, and $\mathbf{p}$ satisfies the corresponding inequality (3). Thus

$$(\mathbf{a}_i+\lambda(\mathbf{q}-\mathbf{a}_i)).(\mathbf{a}_k-\mathbf{a}_i) \leqslant (\tfrac{1}{2}\mathbf{a}_i+\tfrac{1}{2}\mathbf{a}_k).(\mathbf{a}_k-\mathbf{a}_i),$$

or
$$\lambda(\mathbf{q}-\mathbf{a}_i).(\mathbf{a}_k-\mathbf{a}_i) \leqslant \tfrac{1}{2}(\mathbf{a}_k-\mathbf{a}_i).(\mathbf{a}_k-\mathbf{a}_i).$$

Subtracting $(\mathbf{q}-\mathbf{a}_i).(\mathbf{a}_k-\mathbf{a}_i)$ from each side we obtain

$$(\lambda-1)\{(\mathbf{q}-\mathbf{a}_i).(\mathbf{a}_k-\mathbf{a}_i)\} \leqslant \tfrac{1}{2}(\mathbf{a}_k+\mathbf{a}_i-2\mathbf{q}).(\mathbf{a}_k-\mathbf{a}_i),$$

or, on rearranging each side,

$$(\lambda-1)\{(\mathbf{q}-\mathbf{a}_i).(\mathbf{q}-\mathbf{a}_i)-(\mathbf{q}-\mathbf{a}_i).(\mathbf{q}-\mathbf{a}_k)\} \leqslant \tfrac{1}{2}(\mathbf{a}_k-\mathbf{q}).(\mathbf{a}_k-\mathbf{q})-\tfrac{1}{2}(\mathbf{a}_i-\mathbf{q}).(\mathbf{a}_i-\mathbf{q}).$$

Since

$$(\mathbf{q}-\mathbf{a}_i).(\mathbf{q}-\mathbf{a}_i)-(\mathbf{q}-\mathbf{a}_i).(\mathbf{q}-\mathbf{a}_k) \geqslant |\mathbf{q}-\mathbf{a}_i|^2-|\mathbf{q}-\mathbf{a}_i|.|\mathbf{q}-\mathbf{a}_k| > R^2-R^2 = 0,$$

while

$$\tfrac{1}{2}(\mathbf{a}_k-\mathbf{q}).(\mathbf{a}_k-\mathbf{q})-\tfrac{1}{2}(\mathbf{a}_i-\mathbf{q}).(\mathbf{a}_i-\mathbf{q}) = \tfrac{1}{2}|\mathbf{a}_k-\mathbf{q}|^2-\tfrac{1}{2}|\mathbf{a}_i-\mathbf{q}|^2 < \tfrac{1}{2}R^2-\tfrac{1}{2}R^2 = 0,$$

it follows that $\lambda < 1$. Hence

$$|\mathbf{p}-\mathbf{a}_i| = \lambda|\mathbf{q}-\mathbf{a}_i| = \lambda R < R. \tag{4}$$

Consequently, for any integer j not in J_i, we have

$$|\mathbf{p}-\mathbf{a}_j| = |(\mathbf{a}_j-\mathbf{a}_i)-(\mathbf{p}-\mathbf{a}_i)| \geqslant |\mathbf{a}_j-\mathbf{a}_i|-|\mathbf{p}-\mathbf{a}_i| \geqslant 2R-R = R > |\mathbf{p}-\mathbf{a}_i|.$$

Thus the point $\mathbf{p}$ of $\Pi'(\mathbf{a}_i)$ satisfies the inequality (3), not only for the positive integers j in J_i, but also for those not in J_i. Hence $\mathbf{p}$ lies in $\Pi(\mathbf{a}_i)$. As $\mathbf{p}$ was an arbitrary point of the set $\Pi'(\mathbf{a}_i)$, containing $\Pi(\mathbf{a}_i)$, it follows that $\Pi'(\mathbf{a}_i)$ and $\Pi(\mathbf{a}_i)$ coincide.

Since the sequence $\mathbf{a}_1$, $\mathbf{a}_2, \ldots$ is discrete, there are only a finite number of points $\mathbf{a}_j$ satisfying the condition (2). So $\Pi'(\mathbf{a}_i)$, and consequently $\Pi(\mathbf{a}_i)$, being the intersection of a finite number of closed half-spaces, each containing $\mathbf{a}_i$, is a closed convex polyhedron. It follows further, from the inequality (4), that $\Pi(\mathbf{a}_i)$ is contained in the sphere with centre $\mathbf{a}_i$ and radius R.

By our original supposition, corresponding to each point $\mathbf{x}$ of space there will be a positive integer i with

$$|\mathbf{x}-\mathbf{a}_i| < R,$$

but there will only be a finite number of such integers. So it is always possible to choose an integer $i = i(\mathbf{x})$, so that

$$|\mathbf{x}-\mathbf{a}_i|$$

attains its minimum value when $i = i(\mathbf{x})$. Then

$$|\mathbf{x}-\mathbf{a}_{i(\mathbf{x})}| \leqslant |\mathbf{x}-\mathbf{a}_j| \quad (\text{all } j \neq i(\mathbf{x})).$$

Consequently, each point $\mathbf{x}$ lies in at least one of the sets $\Pi(\mathbf{a}_i)$.

If a point $\mathbf{x}$ belongs to two of the sets $\Pi(\mathbf{a}_i)$, say $\Pi(\mathbf{a}_i)$ and $\Pi(\mathbf{a}_j)$, then we have both

$$|\mathbf{x}-\mathbf{a}_i| \leqslant |\mathbf{x}-\mathbf{a}_j|$$

and

$$|\mathbf{x}-\mathbf{a}_j| \leqslant |\mathbf{x}-\mathbf{a}_i|,$$

so that

$$|\mathbf{x}-\mathbf{a}_i| = |\mathbf{x}-\mathbf{a}_j|$$

and

$$\mathbf{x}.(\mathbf{a}_j-\mathbf{a}_i) = (\tfrac{1}{2}\mathbf{a}_i+\tfrac{1}{2}\mathbf{a}_j).(\mathbf{a}_j-\mathbf{a}_i). \tag{5}$$

Consequently $\mathbf{x}$ belongs to one of the $(n-1)$-dimensional faces bounding $\Pi(\mathbf{a}_i)$, and also to one of the $(n-1)$-dimensional faces bounding $\Pi(\mathbf{a}_j)$.

We summarize the results of this section. *Under our assumption concerning the sequence* $\mathbf{a}_1, \mathbf{a}_2, \ldots$, *the set* $\Pi(\mathbf{a}_i)$ *of all points* $\mathbf{x}$, *whose distance from* $\mathbf{a}_i$ *is equal to their minimum distance from the points of* $\{\mathbf{a}_j\}$, *is a closed bounded convex polyhedron. Each point of space belongs to at least one such polyhedron, and can only belong to two or more of the polyhedra if it is common to their boundaries.*

2. The dissection of the Voronoi polyhedra

As in the last section, we suppose that the set of points

$$\mathbf{a}_1, \mathbf{a}_2, \ldots$$

is discrete, and that there is a positive number R, with the property that, for each point $\mathbf{x}$ of space there is a point $\mathbf{a}_i$ of the sequence whose distance from $\mathbf{x}$ is less than R. We suppose further that by the construction described in the last section, a Voronoi polyhedron $\Pi(\mathbf{a}_i)$ is associated with each point $\mathbf{a}_i$ of the sequence. We study one such polyhedron, $\Pi(\mathbf{a})$ say, and we show how it can be dissected into simplices in a certain systematic way.

If $\mathbf{c}_0, \mathbf{c}_1, \ldots, \mathbf{c}_i$ are points, and F is a set, it is convenient to use $\mathbf{c}_0\mathbf{c}_1\ldots\mathbf{c}_iF$ to denote the least convex cover of the set formed by adjoining the points $\mathbf{c}_0, \mathbf{c}_1, \ldots, \mathbf{c}_i$ to the set F. We write

$$\mathbf{c}_0 = \mathbf{a}.$$

Now the polyhedron $\Pi(\mathbf{a})$ can be dissected into the pyramids

$$\mathbf{c}_0 F_{n-1},$$

where F_{n-1} runs over the (closed) $(n-1)$-dimensional faces of $\Pi(\mathbf{a})$, in the sense that these pyramids fit together, without overlapping (they have no inner points in common), and without gaps to form the polyhedron $\Pi(\mathbf{a})$.

We now proceed by induction. We suppose that, for some i with $1 \leqslant i \leqslant n$, the polyhedron $\Pi(\mathbf{a})$ can be dissected into sets of the form

$$\mathbf{c}_0\mathbf{c}_1 \ldots \mathbf{c}_{i-1}F_{n-i},$$

where F_{n-i} is an $(n-i)$-dimensional face, and, for $j = 1, \ldots, i-1$, the point $\mathbf{c}_j$ lies in an $(n-j)$-dimensional face containing $\mathbf{c}_j\mathbf{c}_{j+1}\ldots\mathbf{c}_{i-1}F_{n-i}$, and is the nearest point of this $(n-j)$-dimensional face to $\mathbf{a}$. We confine our attention to one such set

$$\mathbf{c}_0\mathbf{c}_1 \ldots \mathbf{c}_{i-1}F_{n-i}.$$

Take $\mathbf{c}_i$ to be the point of the $(n-i)$-dimensional face F_{n-i} which is nearest to $\mathbf{a}$. In the case when $i = n$, the point $\mathbf{c}_n$ coincides with the 0-dimensional face, or vertex, F_0, and the set

$$\mathbf{c}_0\mathbf{c}_1 \ldots \mathbf{c}_{n-1}F_0$$

becomes the simplex $\quad \mathbf{c}_0\mathbf{c}_1 \ldots \mathbf{c}_n.$

When $i < n$ we note that the sets

$$\mathbf{c}_i F_{n-i-1},$$

where F_{n-i-1} runs over those $(n-i-1)$-dimensional faces that lie F_{n-i} but do not contain $\mathbf{c}_i$, fit together and make up the face F_{n-i}. Since

$$\mathbf{c}_0 \mathbf{c}_1 \ldots \mathbf{c}_{i-1} \mathbf{c}_i F_{n-i-1} = \mathbf{c}_0 \mathbf{c}_1 \ldots \mathbf{c}_{i-1}(\mathbf{c}_i F_{n-i-1}),$$

these sets
$$\mathbf{c}_0 \mathbf{c}_1 \ldots \mathbf{c}_i F_{n-i-1}$$

fit together to form the set

$$\mathbf{c}_0 \mathbf{c}_1 \ldots \mathbf{c}_{i-1} F_{n-i}.$$

Hence, the totality of all the sets of the form

$$\mathbf{c}_0 \mathbf{c}_1 \ldots \mathbf{c}_i F_{n-i-1}$$

fit together to make up the whole polyhedron. Further, in each case F_{n-i-1} is an $(n-i-1)$-dimensional face, and for $j = 1, \ldots, i$, the point $\mathbf{c}_j$ lies in an $(n-j)$-dimensional face containing $\mathbf{c}_j \mathbf{c}_{j+1} \ldots \mathbf{c}_i F_{n-i-1}$, and is the nearest point of this $(n-j)$-dimensional face to $\mathbf{a}$.

It now follows, by induction, that $\Pi(\mathbf{a})$ *can be dissected into simplices of the form*

$$\mathbf{c}_0 \mathbf{c}_1 \ldots \mathbf{c}_n,$$

where $\mathbf{c}_0 = \mathbf{a}$ *and, for* $j = 1, 2, \ldots, n$, *the point* $\mathbf{c}_j$ *lies in an* $(n-j)$-*dimensional face of* $\Pi(\mathbf{a})$ *containing* $\mathbf{c}_j \mathbf{c}_{j+1} \ldots \mathbf{c}_n$, *and is the nearest point of this face to* $\mathbf{a}$.

3. The inequality of Blichfeldt and its application to the Voronoi polyhedra

We show how a simple inequality used by Blichfeldt (1914, 1929) in his work on the packing of spheres can be used to obtain information about the Voronoi polyhedra, and about the simplices into which they can be dissected.

Let K be the sphere in n-dimensional space with radius 1 and with centre $\mathbf{o}$. Let $\mathscr{K}$ be a packing of translates $K + \mathbf{a}_i$ $(i = 1, 2, \ldots)$, where the displacement vectors $\mathbf{a}_1, \mathbf{a}_2 \ldots$ satisfy the condition stated in § 1. We study one of the Voronoi polyhedra, say $\Pi(\mathbf{a})$, and prove:

LEMMA 1. *The distance from* $\mathbf{a}$ *of any point of the* $(n-i)$-*dimensional plane, determined by any* $(n-i)$-*dimensional face of* $\Pi(\mathbf{a})$, *is at least*

$$\sqrt{\left(\frac{2i}{i+1}\right)}$$

for $1 \leqslant i \leqslant n$.

Proof. The $(n-i)$-dimensional plane, determined by any $(n-i)$-dimensional face of $\Pi(\mathbf{a})$, is the intersection of at least i $(n-1)$-dimensional planes, containing $(n-1)$-dimensional faces of $\Pi(\mathbf{a})$. Hence, there are i centres, $\mathbf{a}_1, \mathbf{a}_2, \ldots, \mathbf{a}_i$ say, such that the $(n-i)$-dimensional plane is in the intersection of the $(n-1)$-dimensional planes bisecting the segments $\mathbf{a}\mathbf{a}_1, \mathbf{a}\mathbf{a}_2, \ldots, \mathbf{a}\mathbf{a}_i$. Consider any point $\mathbf{f}$ on the $(n-i)$-dimensional plane. Since $\mathbf{f}$ is equidistant from $\mathbf{a}$ and $\mathbf{a}_1$, from $\mathbf{a}$ and $\mathbf{a}_2 \ldots$, and from $\mathbf{a}$ and $\mathbf{a}_i$, it is clear that $\mathbf{f}$ is equidistant from the points $\mathbf{a}, \mathbf{a}_1, \ldots, \mathbf{a}_i$.

Now, following Blichfeldt, we take $\mathbf{f}$ as origin of a new coordinate system; we also write $\mathbf{a}_0 = \mathbf{a}$, and use the standard scalar product notation. Since none of the spheres of the system overlap, we have a special case of Blichfeldt's inequality

$$\begin{aligned}
2i(i+1) &= \sum_{0 \leqslant \alpha < \beta \leqslant i} 4 \\
&\leqslant \sum_{0 \leqslant \alpha < \beta \leqslant i} (\mathbf{a}_\alpha - \mathbf{a}_\beta).(\mathbf{a}_\alpha - \mathbf{a}_\beta) \\
&= (i+1) \sum_{0 \leqslant \alpha \leqslant i} \mathbf{a}_\alpha.\mathbf{a}_\alpha - \Big(\sum_{0 \leqslant \alpha \leqslant i} \mathbf{a}_\alpha\Big).\Big(\sum_{0 \leqslant \alpha \leqslant i} \mathbf{a}_\alpha\Big) \\
&\leqslant (i+1) \sum_{0 \leqslant \alpha \leqslant i} \mathbf{a}_\alpha.\mathbf{a}_\alpha = (i+1)^2\, \mathbf{a}.\mathbf{a}.
\end{aligned}$$

Thus
$$\mathbf{a}.\mathbf{a} \geqslant \frac{2i}{i+1},$$

nd the distance of the origin $\mathbf{f}$ from the point $\mathbf{a}$ is at least

$$\sqrt{\left(\frac{2i}{i+1}\right)}.$$

This proves the lemma.

LEMMA 2. *If one of the Voronoi polyhedra is dissected into simplices of the form*

$$\mathbf{c}_0\mathbf{c}_1\ldots\mathbf{c}_n$$

as described in §2, *and if* $\mathbf{c}_0$ *is taken as origin of co-ordinates, then*

$$\mathbf{c}_i.\mathbf{c}_j \geqslant \frac{2i}{i+1}$$

for $1 \leqslant i \leqslant j \leqslant n$.

Proof. We suppose that one of the Voronoi polyhedra, corresponding to the packing $\mathscr{K}$, is dissected as described in §2, and that $\mathbf{c}_0\mathbf{c}_1\ldots\mathbf{c}_n$ is one of the resultant simplices. Take $\mathbf{c}_0 = \mathbf{a}$ as the origin. Suppose that $1 \leqslant i \leqslant j \leqslant n$. Then $\mathbf{c}_i$ belongs to an $(n-i)$-dimensional face of $\Pi(\mathbf{a})$, and so

$$\mathbf{c}_i.\mathbf{c}_i \geqslant \frac{2i}{i+1} \tag{6}$$

by Lemma 1. If $j = i$ there is nothing further to prove; so we suppose that $i < j$. Then $\mathbf{c}_i$ is the point of a certain face of $\Pi(\mathbf{a})$, containing $\mathbf{c}_j$, nearest to $\mathbf{a}$. As the faces of $\Pi(\mathbf{a})$ are convex, the whole segment $\mathbf{c}_i\mathbf{c}_j$ belongs to this face. Hence

$$(\mathbf{c}_i+\theta\{\mathbf{c}_j-\mathbf{c}_i\}).(\mathbf{c}_i+\theta\{\mathbf{c}_j-\mathbf{c}_i\}) \geqslant \mathbf{c}_i.\mathbf{c}_i$$

for $0 \leqslant \theta \leqslant 1$. Thus

$$2\theta\mathbf{c}_i.(\mathbf{c}_j-\mathbf{c}_i) \geqslant -\theta^2(\mathbf{c}_j-\mathbf{c}_i).(\mathbf{c}_j-\mathbf{c}_i)$$

for $0 \leqslant \theta \leqslant 1$. As this holds for arbitrarily small positive θ, we deduce that

$$\mathbf{c}_i.(\mathbf{c}_j-\mathbf{c}_i) \geqslant 0.$$

Hence, using (6), we have

$$\mathbf{c}_i.\mathbf{c}_j = \mathbf{c}_i.\mathbf{c}_i+\mathbf{c}_i.(\mathbf{c}_j-\mathbf{c}_i) \geqslant \frac{2i}{i+1}$$

as required.

4. The density of a packing of spheres

Before we state the main theorem of this chapter, we need to describe a certain geometrical configuration. Consider a regular simplex in n-dimensions of side 2, and the system of $n+1$ spheres of radius 1, with their centres at the vertices of the simplex. The spheres of the system do not overlap; let σ_n denote the ratio of the volume of the part of the simplex covered by the spheres to the volume of the whole simplex. In this section we prove

THEOREM 7.1. *If K is an n-dimensional sphere,*

$$\delta(K) \leqslant \sigma_n. \tag{7}$$

Proof. We suppose that

$$\delta(K) > \sigma_n$$

and obtain a contradiction. By Theorem 1.7 we may suppose that K is the sphere with radius 1 and centre $\mathbf{o}$. Again, by Theorem 1.7 we have

$$\delta_P(K) > \sigma_n.$$

Hence we can find a periodic packing $\mathscr{K}_P$ of sets

$$K + \mathbf{a}_i + \mathbf{b}_j \quad (i = 1, 2, \ldots, N;\ j = 1, 2, \ldots),$$

where $\mathbf{b}_1, \mathbf{b}_2, \ldots$ are the points of the lattice Λ of points, whose coordinates are integral multiples of s, such that

$$\rho_+(\mathscr{K}_P) > \sigma_n.$$

Let C be a half-open cube, with its edges parallel to the co-ordinate axes and of edge-length s. Then, by Theorem 1.5,

$$\frac{N\mu(K)}{\mu(C)} = \rho_+(\mathscr{K}_P) > \sigma_n. \tag{8}$$

Now the system of points

$$\mathbf{a}_i + \mathbf{b}_j \quad (i = 1, 2, \ldots, N;\ j = 1, 2, \ldots)$$

certainly satisfies the condition of §1 of this chapter, with

$$R = 2s(C)\sqrt{n}.$$

So we can appeal to the construction of §1 to provide us with a dissection of the whole of space into Voronoi polyhedra

$$\Pi(\mathbf{a}_i + \mathbf{b}_j) \quad (i = 1, 2, \ldots, N;\ j = 1, 2, \ldots).$$

Further, the construction of §2 enables us to dissect each such polyhedron into a finite number of simplices $\mathbf{c}_0\mathbf{c}_1\ldots\mathbf{c}_n$ satisfying the conditions established in §3.

Since each stage of the construction is definitive, the whole geometric structure is periodic in each co-ordinate, with period $s(C)$. So, from among those simplices which have a point in common with C, we can choose a maximal system, $T_1, T_2, \ldots T_M$ say, with the property that no two differ merely by a translation by a vector of the lattice Λ. Then each simplex of the system differs from just one of the simplices $T_1, T_2, \ldots, T_M$ by a translation by a vector of Λ. So the sets

$$T_k + \mathbf{b}_j \quad (k = 1, 2, \ldots, M;\ j = 1, 2, \ldots)$$

fit together to fill the whole of space, the only overlappings being of dimension $n-1$. Hence, we have

$$\begin{aligned} \mu(C) &= \sum_{k=1}^{M} \sum_{j=1}^{\infty} \mu(C \cap \{T_k + \mathbf{b}_j\}) \\ &= \sum_{k=1}^{M} \sum_{j=1}^{\infty} \mu(\{C - \mathbf{b}_j\} \cap T_k) \\ &= \sum_{k=1}^{M} \mu(T_k), \end{aligned} \tag{9}$$

since the cubes $\quad C - \mathbf{b}_j \quad (j = 1, 2, \ldots)$

fit together to fill space exactly. Similarly,

$$\begin{aligned} N\mu(K) &= \sum_{i=1}^{N} \mu(K + \mathbf{a}_i) \\ &= \sum_{i=1}^{N} \sum_{k=1}^{M} \sum_{j=1}^{\infty} \mu(\{K + \mathbf{a}_i\} \cap \{T_k + \mathbf{b}_j\}) \\ &= \sum_{k=1}^{M} \sum_{i=1}^{N} \sum_{j=1}^{\infty} \mu(\{K + \mathbf{a}_i - \mathbf{b}_j\} \cap T_k) \\ &= \sum_{k=1}^{M} \sum_{i=1}^{N} \sum_{j=1}^{\infty} \mu(\{K + \mathbf{a}_i + \mathbf{b}_j\} \cap T_k), \end{aligned}$$

since the vectors $\quad -\mathbf{b}_j \quad (j = 1, 2, \ldots)$

are a rearrangement of the vectors

$$\mathbf{b}_j \quad (j = 1, 2, \ldots).$$

But, owing to the construction of the simplices as subsets of the Voronoi polyhedron, the only sphere $K + \mathbf{a}_i + \mathbf{b}_j$ with any point in common with the simplex T_k, is the sphere centred on its '$\mathbf{c}_0$' vertex, say $\mathbf{c}_0^{(k)}$. Thus

$$N\mu(K) = \sum_{k=1}^{M} \mu(\{K + \mathbf{c}_0^{(k)}\} \cap T_k). \tag{10}$$

It now follows from (8), (9) and (10) that for at least one integer k

$$\mu(\{K + \mathbf{c}_0^{(k)}\} \cap T_k) > \sigma_n \mu(T_k).$$

So, for at least one simplex $T = \mathbf{c}_0 \mathbf{c}_1 \ldots \mathbf{c}_n$ of the system, we have

$$\mu(\{K + \mathbf{c}_0\} \cap T) > \sigma_n \mu(T).$$

We study such a simplex; we may suppose the origin taken so that $\mathbf{c}_0 = \mathbf{o}$. We consider the effect of a certain linear transformation, transforming T into a certain cannonical form.

Consider the n-dimensional regular simplex of side 2, with vertices at the points with coordinates

$$(\sqrt{2}, 0, 0, \ldots, 0),$$
$$(0, \sqrt{2}, 0, \ldots, 0),$$
$$\ldots\ldots\ldots\ldots\ldots\ldots$$
$$(0, 0, 0, \ldots, \sqrt{2})$$

in $(n+1)$-dimensional space. This simplex can be divided into $(n+1)!$ simplices by the following construction. A typical simplex

$$\mathfrak{g}_0 \mathfrak{g}_1 \cdots \mathfrak{g}_n$$

is formed by taking $\mathfrak{g}_0$ to be a vertex, and, when $\mathfrak{g}_0, \mathfrak{g}_1, \ldots, \mathfrak{g}_{i-1}$ have been chosen, taking $\mathfrak{g}_i$ to be the centroid of one of the $n-i+1$ i-dimensional faces of the simplex that contains the points $\mathfrak{g}_0, \mathfrak{g}_1, \ldots, \mathfrak{g}_{i-1}$. In particular, one of these simplices G is obtained by taking

$$\mathfrak{g}_i = \left(\frac{\sqrt{2}}{i+1}, \ldots, \frac{\sqrt{2}}{i+1}, 0, \ldots, 0\right)$$

for $i = 0, 1, 2, \ldots, n$, where $i+1$ of the coordinates are non-zero. Now it is easy to verify that, if $1 \leqslant i \leqslant j \leqslant n$,

$$(\mathfrak{g}_i - \mathfrak{g}_0) \cdot (\mathfrak{g}_j - \mathfrak{g}_0) = \frac{2ij}{(i+1)(j+1)} + \frac{2i}{(i+1)(j+1)} = \frac{2i}{i+1}.$$

Hence, by Lemma 2 of §3,

$$\mathbf{c}_i.\mathbf{c}_j \geqslant (\mathfrak{g}_i-\mathfrak{g}_0).(\mathfrak{g}_i-\mathfrak{g}_0)$$

if $1 \leqslant i \leqslant j \leqslant n$.

Now consider the linear transformation L, transforming the general point

$$\lambda_1\mathbf{c}_1+\lambda_2\mathbf{c}_2+\ldots+\lambda_n\mathbf{c}_n$$

into the corresponding point

$$\mathfrak{g}_0+\sum_{i=1}^{n}\lambda_i(\mathfrak{g}_i-\mathfrak{g}_0)$$

of the n-dimensional space of points $\mathbf{y}$ with

$$y_1+y_2+\ldots+y_{n+1}=\sqrt{2}.$$

Clearly $\mathsf{L}T = G$. Also $\mathsf{L}K$ is a certain ellipsoid lying in the same n-dimensional space as G. Now, if $\mathbf{x}$ is a point of T in K, we have

$$\mathbf{x}=\sum_{i=1}^{n}\lambda_i\mathbf{c}_i,$$

where $\lambda_i \geqslant 0$ $(i = 1, 2, \ldots, n)$ and

$$\mathbf{x}.\mathbf{x} \leqslant 1.$$

Hence, the corresponding point $\mathbf{y}$ of $E \cap G$ satisfies

$$\begin{aligned}(\mathbf{y}-\mathfrak{g}_0).(\mathbf{y}-\mathfrak{g}_0)&\\ &=\sum_{i=1}^{n}\sum_{j=1}^{n}\lambda_i\lambda_j(\mathfrak{g}_i-\mathfrak{g}_0).(\mathfrak{g}_j-\mathfrak{g}_0)\\ &\leqslant\sum_{i=1}^{n}\sum_{j=1}^{n}\lambda_i\lambda_j\mathbf{c}_i.\mathbf{c}_j\\ &=\mathbf{x}.\mathbf{x}\leqslant 1.\end{aligned}$$

Thus $E \cap G$ is contained in $S \cap G$, where S is the sphere with centre $\mathfrak{g}_0$ and radius 1.

But, as L preserves the ratio of two volumes, we have

$$\sigma_n < \frac{\mu(K \cap T)}{\mu(T)} = \frac{\mu(E \cap G)}{\mu(G)} \leqslant \frac{\mu(S \cap G)}{\mu(G)}.$$

But it is clear from the definition of σ_n, and from the congruence of the simplices into which the regular simplex of side 2 is divided, that

$$\frac{\mu(S \cap G)}{\mu(G)} = \sigma_n.$$

This contradiction proves the theorem.

5. Daniels's asymptotic formula

In this section, we derive an asymptotic formula, due to H. E. Daniels (see Rogers, 1958*c*), for the ratio σ_n introduced in the last section. We use a slight modification of his method.

By definition, σ_n is the ratio of the volume of that part of a regular n-dimensional simplex of side 2 covered by the system of $n+1$ spheres of radius 1, with their centres at the vertices of the simplex, to the volume of the whole simplex. Since the spheres of the system do not overlap, it follows, from considerations of symmetry and homogeneity, that

$$\sigma_n = (n+1)\rho_n,$$

where ρ_n is the ratio of the volume of that part of a regular simplex of side $\sqrt{2}$ covered by a sphere having radius $1/\sqrt{2}$ and its centre at a vertex, to the volume of the whole simplex. In particular, we may consider the regular simplex in n-dimensional space, with vertices

$$(1, 0, \ldots, 0),$$
$$(0, 1, \ldots, 0),$$
$$\ldots\ldots\ldots\ldots\ldots\ldots$$
$$(0, 0, \ldots, 1),$$
$$(-a, -a, \ldots, -a),$$

where a is the positive root of the equation

$$na^2 + 2a = 1.$$

Then we can take the sphere defined by the inequality

$$(x_1+a)^2 + \ldots + (x_n+a)^2 \leqslant \tfrac{1}{2}.$$

We apply the linear transformation given by

$$y_i = (x_i+a) - \frac{a}{1+na}\sum_{j=1}^{n}(x_j+a) \quad (i = 1, 2, \ldots, n),$$

or by
$$x_i = y_i - a + a\sum_{j=1}^{n} y_j \quad (i = 1, 2, \ldots, n)$$

to the simplex and the sphere. The simplex transforms into the simplex with vertices

$$(1, 0, \ldots, 0),$$

$$(0, 1, \ldots, 0),$$

$$\ldots\ldots\ldots\ldots\ldots$$

$$(0, 0, \ldots, 1),$$

$$(0, 0, \ldots, 0),$$

and volume $1/n!$. The sphere transforms into the ellipsoid given by the inequality

$$\sum_{j=1}^{n} y_j^2 + \left(\sum_{j=1}^{n} y_j\right)^2 \leqslant \tfrac{1}{2}.$$

It follows immediately that

$$\sigma_n = (n+1)\rho_n = (\tfrac{1}{2})^{\frac{1}{2}n}[(n+1)!]V_n,$$

where V_n is the volume of the region defined by the inequalities

$$\sum_{j=1}^{n} y_j^2 + \left(\sum_{j=1}^{n} y_j\right)^2 \leqslant 1,$$

$$y_1 \geqslant 0, \quad y_2 \geqslant 0, \ldots, y_n \geqslant 0.$$

By homogeneity, the volume of the sector $S(Y)$ of the ellipsoid

$$\sum_{j=1}^{n} y_j^2 + \left(\sum_{j=1}^{n} y_j\right)^2 \leqslant Y$$

lying in the positive octant is $Y^{\frac{1}{2}n}V_n$. Writing

$$F(y_1, \ldots, y_n) = \exp\left[-\sum_{j=1}^{n} y_j^2 - \left(\sum_{j=1}^{n} y_j\right)^2\right],$$

we have

$$\int_0^\infty \int_0^\infty \cdots \int_0^\infty \exp\left[-\sum_{j=1}^n y_j^2 - \left(\sum_{j=1}^n y_j\right)^2\right] dy_1 dy_2 \ldots dy_n$$

$$= \int_0^\infty \int_0^\infty \cdots \int_0^\infty \left\{\int_0^{F(y_1, \ldots, y_n)} dZ\right\} dy_1 dy_2 \ldots dy_n$$

$$= \int_0^1 \left\{\underset{F(y_1, \ldots, y_n) \geqslant Z}{\int_0^\infty \int_0^\infty \cdots \int_0^\infty} dy_1 dy_2 \ldots dy_n\right\} dZ$$

$$= \int_0^1 \mu(S(\log 1/Z))\, dZ$$

$$= \int_0^1 (\log 1/Z)^{n/2} V_n\, dZ$$

$$= V_n \Gamma(1 + \tfrac{1}{2}n).$$

Now if in the elementary result

$$\sqrt{\pi} = \int_{-\infty}^{\infty} e^{-w^2} dw,$$

we regard the right-hand side as a contour integral taken along the real axis, it is easy to see that we shall obtain the same value if we integrate along the straight line

$$w = u - iy, \quad -\infty < u < +\infty,$$

where y is a real constant and u a real parameter. Hence

$$\sqrt{\pi} = \int_{-\infty}^{\infty} e^{-u^2 + 2iuy + y^2} du,$$

so that

$$(\sqrt{\pi})\, e^{-y^2} = \int_{-\infty}^{\infty} e^{-w^2 + 2iwy} dw.$$

Hence, using this with $y = \Sigma y_j$,

$$2^{n/2}(\sqrt{\pi})\, \Gamma(1 + \tfrac{1}{2}n)\, \sigma_n/(n+1)!$$

$$= (\sqrt{\pi})\, \Gamma(1 + \tfrac{1}{2}n)\, V_n$$

$$= \int_0^\infty \int_0^\infty \cdots \int_0^\infty \left\{\int_{-\infty}^{\infty} \exp\left[-\sum_{j=1}^n y_j^2 - w^2 + 2iw \sum_{j=1}^n y_j\right] dw\right\}$$

$$\times dy_1 dy_2 \ldots dy_n.$$

Thus, changing the order of the integrations, as is permissible, since the integrand is absolutely integrable, we have

$$\frac{2^{\frac{1}{2}n}(\sqrt{\pi})\,\Gamma(1+\frac{1}{2}n)\,\sigma_n}{(n+1)!}$$

$$= \int_{-\infty}^{\infty}\left\{\int_0^{\infty}\int_0^{\infty}\cdots\int_0^{\infty} e^{-w^2}\prod_{j=1}^{n}\exp(-y_j^2+2iwy_j)\right.$$

$$\left.\times\, dy_1 dy_2 \ldots dy_n\right\} dw$$

$$= \int_{-\infty}^{\infty} e^{-w^2}\left[\int_0^{\infty} e^{-y^2+2iwy}\,dy\right]^n dw.$$

Now the integrand

$$f(w) = e^{-w^2}\left[\int_0^{\infty} e^{-y^2+2iwy}\,dy\right]^n,$$

regarded as a function of the complex variable w, is an integral function. Further, if $w = u+iv$, with u, v real, we have

$$|f(w)| \leqslant e^{-u^2+v^2}\left[\int_0^{\infty} e^{-y^2}\,dy\right]^n$$

$$= (\tfrac{1}{2}\sqrt{\pi})^n\, e^{-u^2+v^2}$$

for $v \geqslant 0$. Thus $f(w)$ tends rapidly to zero, as w tends to infinity in any strip in the upper half-plane bounded by lines parallel to the real axis. Hence, we can easily show that we can replace the integration along the real axis by integration along the line given parametrically by

$$w = u+i\nu, \quad -\infty < u < +\infty,$$

where we take $\nu = (\frac{1}{2}n)^{\frac{1}{2}}$. This particular contour is chosen as a convenient approximation to a path of steepest descent, through the saddle-point of the function $f(w)$ on the imaginary axis. We obtain

$$\frac{2^{\frac{1}{2}n}(\sqrt{\pi})\,\Gamma(1+\frac{1}{2}n)\,\sigma_n}{(n+1)!}$$

$$= \int_{-\infty}^{\infty} e^{-u^2-2i\nu u+\nu^2}\left[\int_0^{\infty} e^{-y^2-2\nu y+2iuy}\,dy\right]^n du$$

$$= \frac{e^{\frac{1}{2}n}}{(2n)^{\frac{1}{2}n}}\int_{-\infty}^{\infty} e^{-u^2}\left[2\nu\, e^{-iu/\nu}\int_0^{\infty} e^{-y^2-2\nu y+2iuy}\,dy\right]^n du.$$

Since
$$\left|2\nu e^{-iu/\nu}\int_0^\infty e^{-y^2-2\nu y+2iuy}\,dy\right|$$
$$\leqslant 2\nu\int_0^\infty e^{-y^2-2\nu y}\,dy$$
$$= 2\nu\left\{\frac{1}{2\nu}-\frac{1}{\nu}\int_0^\infty y\,e^{-y^2}e^{-2\nu y}\,dy\right\}$$
$$< 1,$$
the integrand
$$g_n(u) = e^{-u^2}\left[2\nu e^{-iu/\nu}\int_0^\infty e^{-y^2-2\nu y+2iuy}\,dy\right]^n$$
is dominated by the integrable function e^{-u^2} uniformly for all n. So, by the theory of dominated convergence,
$$\lim_{n\to\infty}\int_{-\infty}^\infty g_n(u)\,du = \int_{-\infty}^\infty g(u)\,du,$$
provided we can find a function $g(u)$ such that
$$\lim_{n\to\infty} g_n(u) = g(u).$$

Now, integrating by parts four times, we have
$$\int_0^\infty e^{-y^2}e^{-\mu y}\,dy = \frac{1}{\mu}-\frac{2}{\mu^3}+\frac{1}{\mu^4}\int_0^\infty e^{-\mu y}\left(\frac{d^4}{dy^4}e^{-y^2}\right)dy.$$
Provided the real part of μ is positive, and its imaginary part is bounded, on integrating by parts once more we have
$$\int_0^\infty e^{-y^2}e^{-\mu y}\,dy = \frac{1}{\mu}-\frac{2}{\mu^3}+O\left(\frac{1}{|\mu|^5}\right),$$
as $|\mu|\to\infty$. Taking
$$\mu = 2\nu-2iu,\quad \nu = (\tfrac{1}{2}n)^{\frac{1}{2}},$$
we have
$$2\nu e^{-iu/\nu}\int_0^\infty e^{-y^2-2\nu y+2iuy}\,dy$$
$$= 2\nu\left(1-\frac{iu}{\nu}-\frac{u^2}{n}+O\left(\frac{1}{n^{\frac{3}{2}}}\right)\right)$$
$$\times\left(\frac{1}{2\nu-2iu}-\frac{2}{(2\nu-2iu)^3}+O\left(\frac{1}{n^{\frac{5}{2}}}\right)\right)$$
$$= 1-\frac{u^2}{n}-\frac{1}{n}+O\left(\frac{1}{n^{\frac{3}{2}}}\right).$$

Consequently, for each fixed u

$$\lim_{n\to\infty} g_n(u) = \lim_{n\to\infty} e^{-u^2}\left[2\nu\, e^{-iu/\nu}\int_0^\infty e^{-y^2-2\nu y+2iuy}\,dy\right]^n$$
$$= e^{-2u^2-1}.$$

Hence

$$\frac{2^{\frac{1}{2}n}(\sqrt{\pi})\,\Gamma(1+\tfrac{1}{2}n)\,\sigma_n}{(n+1)!}$$
$$= \frac{e^{\frac{1}{2}n}}{(2n)^{\frac{1}{2}n}}\int_{-\infty}^{\infty} g_n(u)\,du$$
$$\sim \frac{e^{\frac{1}{2}n}}{(2n)^{\frac{1}{2}n}}\int_{-\infty}^{\infty} e^{-2u^2-1}\,du$$
$$= \frac{\sqrt{\pi}\, e^{-(n/2)-1}}{(\sqrt{2})\,(2n)^{\frac{1}{2}n}}.$$

So, using Stirling's formula,

$$\sigma_n \sim \frac{(n+1)!\, e^{(n/2)-1}}{(\sqrt{2})\,\Gamma(1+\tfrac{1}{2}n)\,(4n)^{n/2}} \sim \frac{n}{e}\left(\frac{1}{\sqrt{2}}\right)^n. \tag{11}$$

This is Daniels's asymptotic formula.

CHAPTER 8

COVERINGS WITH SPHERES CANNOT BE VERY ECONOMICAL

1. The dual subdivision

In this section we describe a construction, given by Delaunay in 1934, of a certain subdivision of space which is in some ways dual to the subdivision into Voronoi polyhedra. As in §§ 1 and 2 of Chapter 7, we suppose, throughout this section, that *the set of points*

$$\mathbf{a}_1, \mathbf{a}_2, \ldots$$

is discrete, and that there is a positive number R with the property that, for each point $\mathbf{x}$ of space, there is a point $\mathbf{a}_i$ of the sequence whose distance from $\mathbf{x}$ is less than R. We suppose further† that *there is no point of space which is equidistant from $n+2$ points of the sequence, all within distance R.*

With each point $\mathbf{a}_i$ we associate the corresponding Voronoi polyhedron $\Pi(\mathbf{a}_i)$ in the way described in § 1 of Chapter 7. As we noted there, $\Pi(\mathbf{a}_i)$ lies within the sphere with centre $\mathbf{a}_i$ and radius R. By the second assumption, italicized above, each vertex of each Voronoi polyhedron is equidistant from precisely $n+1$ of the points of the sequence, and so is common to precisely $n+1$ of the Voronoi polyhedra. Let V be the set of all vertices of all the Voronoi polyhedra.

Corresponding to the subdivision of space into the Voronoi polyhedra, we have the well-known dual graph obtained by joining those pairs of points $\mathbf{a}_i$, $\mathbf{a}_j$ whose Voronoi polyhedra have an $(n-1)$-dimensional face in common. While the dual graph has been usually considered as an abstract graph expressing the topological relationships of the polytopes, we are here concerned to take the joins to be straight line segments, and to use them as a framework for a simplicial subdivision of space.

With each point $\mathbf{v}$ of V we associate the closed simplex

† See Coxeter, Few and Rogers (1959) for a discussion of the modifications necessary when this further condition is not fulfilled.

$T(\mathbf{v})$, with its vertices at the $n+1$ points of the sequence $\mathbf{a}_1, \mathbf{a}_2, \ldots$, whose Voronoi polyhedra contain $\mathbf{v}$. If these $n+1$ points, say $\mathbf{a}_1, \mathbf{a}_2, \ldots, \mathbf{a}_{n+1}$, were all in the same plane, then each point $\mathbf{l}$, on the line l through $\mathbf{v}$ perpendicular to the plane, would be equidistant from the $n+1$ points $\mathbf{a}_1, \mathbf{a}_2, \ldots, \mathbf{a}_{n+1}$. When $\mathbf{l}$ is at $\mathbf{v}$, it is within a distance R of $\mathbf{a}_1, \mathbf{a}_2, \ldots, \mathbf{a}_{n+1}$. Also, each point $\mathbf{l}$ of l is within a distance less than R of some point $\mathbf{a}_i$ of the sequence. Hence, as $\mathbf{l}$ moves along l from the initial position $\mathbf{v}$, it must reach a position when it is equidistant from

$$\mathbf{a}_1, \mathbf{a}_2, \ldots, \mathbf{a}_{n+1},$$

and an $(n+2)$nd point $\mathbf{a}_j$, while still within a distance R of these points. This is contrary to our suppositions, and shows that the simplex $T(\mathbf{v})$ is necessarily non-degenerate. We shall show that these simplices fit together without overlapping, to cover the whole of space; they will be called the Delaunay simplices.

First, suppose that $T(\mathbf{v})$, $T(\mathbf{w})$ are simplices of the system, corresponding to distinct points $\mathbf{v}$, $\mathbf{w}$ of V. Then, by the construction of the Voronoi polyhedra, and the choice of $\mathbf{v}$ and $T(\mathbf{v})$, the vertices of $T(\mathbf{v})$ lie on a certain sphere $\Sigma(\mathbf{v})$ with centre $\mathbf{v}$, and no point $\mathbf{a}_i$ lies in the interior of $\Sigma(\mathbf{v})$. Thus the vertices of $T(\mathbf{w})$ lie on a corresponding sphere $\Sigma(\mathbf{w})$, but do not lie in the interior of $\Sigma(\mathbf{v})$. Similarly, the vertices of $T(\mathbf{v})$ lie on $\Sigma(\mathbf{v})$, but not in the interior of $\Sigma(\mathbf{w})$. It follows that neither sphere can be contained in the interior of the other, nor can they coincide, as $\mathbf{v} \neq \mathbf{w}$. Further, the vertices of $T(\mathbf{v})$ must be separated from those of $T(\mathbf{w})$ by the radical plane of the two spheres, in the sense that the vertices of $T(\mathbf{v})$ lie in the closed half-space bounded by the radical plane and containing $\mathbf{v}$, while the vertices of $T(\mathbf{w})$ lie in the other closed half-space bounded by the radical plane. Hence the simplices $T(\mathbf{v})$ and $T(\mathbf{w})$ are separated by the radical plane in the same way, and so they contain no common inner points. Thus the simplices of the system are non-overlapping.

Now suppose that there is a point $\mathbf{x}$ of space which belongs to none of the simplices. If $T(\mathbf{v})$ is a simplex of the system, having a point within unit distance of $\mathbf{x}$, then $\mathbf{v}$ lies within distance $R+1$ of $\mathbf{x}$ and is the vertex of a Voronoi polyhedron, $\Pi(\mathbf{a}_i)$

say, with $\mathbf{a}_i$ within a distance $2R+1$ of $\mathbf{x}$. But there are only a finite number of such points $\mathbf{a}_i$, and the corresponding polyhedra have only a finite number of vertices. Thus there are only a finite number of simplices of the system having any points within unit distance of $\mathbf{x}$. Since the simplices are closed, it follows that there is a small sphere Σ with centre $\mathbf{x}$ which contains no point of any simplex. Choose any point $\mathbf{s}$ in the interior of any simplex of the system. Then the set of all points, on the lines joining $\mathbf{s}$ to the $(n-2)$-dimensional faces of all the simplices, is of dimension $n-1$. So we can choose a point $\mathbf{y}$ of Σ, such that the segment $\mathbf{sy}$ meets no $(n-2)$-dimensional face of any of the simplices. Now the segment $\mathbf{sy}$ meets the simplices of the system in a finite number of closed intervals, one of which includes $\mathbf{s}$, but none of which includes $\mathbf{y}$. Hence, if we take $\mathbf{t}$ to be the closed interval end-point nearest to $\mathbf{y}$, the point $\mathbf{t}$ belongs to one of the simplices of the system, say $T(\mathbf{v})$, but no other point of the segment $\mathbf{ty}$ belongs to any simplex of the system. Since $\mathbf{t}$ is on the boundary of $T(\mathbf{v})$, but is not on any $(n-2)$-dimensional face of $T(\mathbf{v})$, it is an inner point of one of the $(n-1)$-dimensional faces of $T(\mathbf{v})$. Let $\mathbf{a}_0, \mathbf{a}_1, \ldots, \mathbf{a}_n$ be the vertices of $T(\mathbf{v})$, named so that $\mathbf{t}$ is an inner point of the simplex $\mathbf{a}_1\mathbf{a}_2\ldots\mathbf{a}_n$. Then the locus of points $\mathbf{x}$, which are equidistant from the points $\mathbf{a}_1, \mathbf{a}_2, \ldots, \mathbf{a}_n$, is the line l, through $\mathbf{v}$, perpendicular to the plane π of the simplex $\mathbf{a}_1\mathbf{a}_2\ldots\mathbf{a}_n$. Let us take the direction of the perpendicular from $\mathbf{a}_0$ to π as the positive direction of l. Then all points of l on the positive side of $\mathbf{v}$, which are sufficiently close to $\mathbf{v}$, are equidistant from $\mathbf{a}_1, \mathbf{a}_2 \ldots, \mathbf{a}_n$, but are at strictly greater distances from all other points $\mathbf{a}_i$. Thus the points of l, just on the positive side of $\mathbf{v}$, lie on each of the $(n-1)$-dimensional faces of $\Pi(\mathbf{a}_1)$, lying in the planes bisecting the segments $\mathbf{a}_1\mathbf{a}_2, \ldots, \mathbf{a}_1\mathbf{a}_n$. Consequently, these points lie on a 1-dimensional face or 'edge' of $\Pi(\mathbf{a}_1)$. This edge is clearly common to the polyhedra

$$\Pi(\mathbf{a}_1), \Pi(\mathbf{a}_2), \ldots, \Pi(\mathbf{a}_n).$$

The vertex $\mathbf{v}$ forms one end of this edge; let $\mathbf{w}$ denote the vertex forming the other end of the edge. Then the simplex $T(\mathbf{w})$, associated with $\mathbf{w}$, has $\mathbf{a}_1, \mathbf{a}_2, \ldots, \mathbf{a}_n$, for n of its vertices; let $\mathbf{a}_{n+1}$ be the $(n+1)$st vertex. Since $T(\mathbf{v})$ and $T(\mathbf{w})$ have no inner points in common, it follows immediately that $\mathbf{a}_0$ and $\mathbf{a}_{n+1}$ are on opposite sides of the plane π of the simplex $\mathbf{a}_1\mathbf{a}_2, \dots \mathbf{a}_n$. Consequently, each inner point of the simplex $\mathbf{a}_1\mathbf{a}_2 \dots \mathbf{a}_n$, and in particular $\mathbf{t}$, is an inner point of the union of $T(\mathbf{v})$ and $T(\mathbf{w})$. Hence, points of the segment $\mathbf{ty}$, other than $\mathbf{t}$, belong to either $T(\mathbf{v})$ or $T(\mathbf{w})$. This contradiction shows that our original supposition is false, and each point of space necessarily belongs to at least one of the simplices. This completes the proof of the main result of this section: *the Delaunay simplices* $T(\mathbf{v})$, *with* $\mathbf{v}$ *in* V, *fit together without overlapping to cover the whole of space.*

2. The solid angles of a simplex

The solid angle of a cone in n-dimensional space is defined to be the surface area of the region cut from the surface of the sphere, with radius 1 and its centre at the vertex of the cone, by the cone. The solid angle of a simplex $\mathbf{a}_0\mathbf{a}_1 \dots \mathbf{a}_n$, at the vertex $\mathbf{a}_0$, is naturally defined as the solid angle of the cone consisting of all points of the form

$$\mathbf{a}_0 + \sum_{i=1}^{n} \lambda_i(\mathbf{a}_i - \mathbf{a}_0),$$

where $\lambda_1, \lambda_2, \dots, \lambda_n$ are non-negative, but do not necessarily satisfy the further restriction that their sum is less than or equal to 1. In this section we prove a lemma, concerning the sum of the solid angles of a simplex, which will be crucial for the proof of our main result in the next section.

LEMMA 1. *Let* T *be an* n*-dimensional simplex contained in a sphere of radius* 1. *Then the ratio* $\Sigma(T)/\mu(T)$ *of the sum* $\Sigma(T)$ *of the vertex solid angles of* T *to the volume of* T *attains its minimum value when* T *is the regular simplex inscribed in the sphere.*

Proof. Let T be the simplex with vertices $\mathbf{a}_0, \mathbf{a}_1, \dots, \mathbf{a}_n$, with coordinates given by

$$\mathbf{a}_i = (a_1^{(i)}, a_2^{(i)}, \dots, a_n^{(i)}) \quad (i = 0, 1, \dots, n).$$

We suppose temporarily that $\mathbf{a}_0$ coincides with the origin $\mathbf{o}$. We introduce new coordinates $(p_1, p_2, \dots, p_n)$ related to the original coordinates $(x_1, x_2 \dots, x_n)$ by the equations

$$x_j = \sum_{\lambda=1}^{n} a_j^{(\lambda)} p_\lambda \quad (i = 1, 2, \dots, n).$$

Then the new coordinates of the vertices of T are

$$(0, 0, \ldots, 0),$$
$$(1, 0, \ldots, 0),$$
$$\ldots\ldots\ldots\ldots\ldots$$
$$(0, 0, \ldots, 1);$$

and T is the set of points $\mathbf{p} = (p_1, p_2, \ldots, p_n)$ satisfying

$$p_1 \geqslant 0, \quad p_2 \geqslant 0, \ldots, p_n \geqslant 0, \quad p_1 + p_2 + \ldots + p_n \leqslant 1.$$

Now the surface area of the unit sphere with centre $\mathbf{a}_0 = \mathbf{o}$ is

$$\frac{2\pi^{\frac{1}{2}n}}{\Gamma(\frac{1}{2}n)}.$$

Further, if the integral is taken over the whole of space,

$$\int e^{-\mathbf{x}.\mathbf{x}} d\mathbf{x} = \left\{\int_{-\infty}^{\infty} e^{-x^2} dx\right\}^n = \pi^{\frac{1}{2}n}.$$

Hence the solid angle Σ_0 of T at the vertex $\mathbf{a}_0$ is given by

$$\Sigma_0 \Big/ \left(\frac{2\pi^{\frac{1}{2}n}}{\Gamma(\frac{1}{2}n)}\right) = \int_{p_i \geqslant 0} e^{-\mathbf{x}.\mathbf{x}} d\mathbf{x} \Big/ \int e^{-\mathbf{x}.\mathbf{x}} d\mathbf{x}$$
$$= \frac{1}{\pi^{\frac{1}{2}n}} \int_{p_i \geqslant 0} e^{-\mathbf{x}.\mathbf{x}} d\mathbf{x}$$
$$= \frac{|\Delta|}{\pi^{\frac{1}{2}n}} \int_0^{\infty} \int_0^{\infty} \ldots \int_0^{\infty} e^{-\mathbf{x}.\mathbf{x}} dp_1 dp_2 \ldots dp_n,$$

where $$\Delta = \det(a_i^{(j)}).$$

Since $$\mathbf{x}.\mathbf{x} = \left(\sum_{\lambda=1}^{n} \mathbf{a}_\lambda p_\lambda\right).\left(\sum_{\lambda=1}^{n} \mathbf{a}_\lambda p_\lambda\right)$$
$$= \sum_{\lambda, \mu=1}^{n} \mathbf{a}_\lambda . \mathbf{a}_\mu p_\lambda p_\mu$$
$$= \sum_{\lambda, \mu=1}^{n} (\mathbf{a}_\lambda - \mathbf{a}_0).(\mathbf{a}_\mu - \mathbf{a}_0) p_\lambda p_\mu,$$

and since the volume of T is

$$\mu(T) = \Delta n!$$

we deduce that

$$\Sigma_0/\mu(T) = \frac{2(n!)}{\Gamma(\frac{1}{2}n)} \int_0^\infty \int_0^\infty \dots \int_0^\infty \exp\left[-\sum_{\lambda,\mu=1}^{n} (\mathbf{a}_\lambda - \mathbf{a}_0).(\mathbf{a}_\mu - \mathbf{a}_0) q_\lambda q_\mu\right] \times dq_1 dq_2 \dots dq_n.$$

Clearly in this formula the assumption that $\mathbf{a}_0$ coincides with $\mathbf{o}$ is no longer relevant.

More generally, the solid angle Σ_j of T, at the vertex $\mathbf{a}_j$, is given for $j = 0, 1, \dots, n$ by

$$\Sigma_j/\mu(T) = \frac{2(n!)}{\Gamma(\frac{1}{2}n)} \int_0^\infty \int_0^\infty \dots \int_0^\infty \exp\left[-\sum_{\lambda,\mu=1}^{n} (\mathbf{a}_{\sigma(\lambda)} - \mathbf{a}_{\sigma(0)}) \times (\mathbf{a}_{\sigma(\mu)} - \mathbf{a}_{\sigma(0)}) q_\lambda q_\mu\right] dq_1 dq_2 \dots dq_n,$$

where $\sigma(0), \sigma(1), \dots, \sigma(n)$ is any one of the $n!$ permutations of the integers $0, 1, \dots, n$ with $\sigma(0) = j$. Summing this equation over all permutations σ of $0, 1, \dots, n$, we obtain the identity

$$n!\,\Sigma(T)/\mu(T) = \frac{2(n!)}{\Gamma(\frac{1}{2}n)} \int_0^\infty \int_0^\infty \dots \int_0^\infty \sum_\sigma \exp\left[-\sum_{\lambda,\mu=1}^{n} (\mathbf{a}_{\sigma(\lambda)} - \mathbf{a}_{\sigma(0)}) \times (\mathbf{a}_{\sigma(\mu)} - \mathbf{a}_{\sigma(0)}) q_\lambda q_\mu\right] dq_1 dq_2 \dots dq_n.$$

Applying the inequality of the arithmetic and geometric means to the integrand we have

$$\Sigma(T)/\mu(T) \geqslant \frac{2}{\Gamma(\frac{1}{2}n)} \int_0^\infty \int_0^\infty \dots \int_0^\infty (n+1)! \exp\left[-\frac{1}{(n+1)!} \sum_\sigma \sum_{\lambda,\mu=1}^{n} \times (\mathbf{a}_{\sigma(\lambda)} - \mathbf{a}_{\sigma(0)}).(\mathbf{a}_{\sigma(\mu)} - \mathbf{a}_{\sigma(0)}) q_\lambda q_\mu\right] dq_1 dq_2 \dots dq_n,$$

with equality holding in the special case, when T is a regular simplex. We suppose that T is contained in the sphere with centre $\mathbf{o}$ and radius 1. Then, when $\lambda \neq \mu$,

$$\frac{1}{(n+1)!}\sum_{\sigma}(\mathbf{a}_{\sigma(\lambda)}-\mathbf{a}_{\sigma(0)}).(\mathbf{a}_{\sigma(\mu)}-\mathbf{a}_{\sigma(0)})$$

$$=\frac{1}{(n+1)!}\{\sum_{\sigma}\mathbf{a}_{\sigma(\lambda)}.\mathbf{a}_{\sigma(\mu)}-\sum_{\sigma}\mathbf{a}_{\sigma(\lambda)}.\mathbf{a}_{\sigma(0)}$$

$$-\sum_{\sigma}\mathbf{a}_{\sigma(0)}.\mathbf{a}_{\sigma(\mu)}+\sum_{\sigma}\mathbf{a}_{\sigma(0)}.\mathbf{a}_{\sigma(0)}\}$$

$$=\frac{1}{(n+1)!}\{(n-1)!\sum_{\alpha\neq\beta}\mathbf{a}_{\alpha}.\mathbf{a}_{\beta}-2\{(n-1)!\}\sum_{\alpha\neq\beta}\mathbf{a}_{\alpha}.\mathbf{a}_{\beta}+n!\sum_{\alpha}\mathbf{a}_{\alpha}.\mathbf{a}_{\alpha}\}$$

$$=\frac{1}{n(n+1)}\left\{(n+1)\sum_{\alpha=0}^{n}\mathbf{a}_{\alpha}.\mathbf{a}_{\alpha}-\left(\sum_{\alpha=0}^{n}\mathbf{a}_{\alpha}\right).\left(\sum_{\beta=0}^{n}\mathbf{a}_{\beta}\right)\right\}$$

$$\leqslant\frac{n+1}{n},$$

with equality, when T is inscribed in the sphere with centre $\mathbf{o}$ and radius 1, and the 'centre of gravity' of T coincides with $\mathbf{o}$. Similarly, when $\lambda = \mu$,

$$\frac{1}{(n+1)!}\sum_{\sigma}(\mathbf{a}_{\sigma(\lambda)}-\mathbf{a}_{\sigma(0)}).(\mathbf{a}_{\sigma(\lambda)}-\mathbf{a}_{\sigma(0)})$$

$$=\frac{1}{(n+1)!}\{\sum_{\sigma}\mathbf{a}_{\sigma(\lambda)}.\mathbf{a}_{\sigma(\lambda)}-2\sum_{\sigma}\mathbf{a}_{\sigma(\lambda)}.\mathbf{a}_{\sigma(0)}+\sum_{\sigma}\mathbf{a}_{\sigma(0)}.\mathbf{a}_{\sigma(0)}\}$$

$$=\frac{1}{(n+1)!}\{2(n!)\sum_{\alpha}\mathbf{a}_{\alpha}.\mathbf{a}_{\alpha}-2\{(n-1)!\}\sum_{\alpha\neq\beta}\mathbf{a}_{\alpha}.\mathbf{a}_{\beta}\}$$

$$=\frac{2}{n(n+1)}\left\{(n+1)\sum_{\alpha=0}^{n}\mathbf{a}_{\alpha}.\mathbf{a}_{\alpha}-\left(\sum_{\alpha=\beta}^{n}\mathbf{a}_{\alpha}\right).\left(\sum_{\beta=0}^{n}\mathbf{a}_{\beta}\right)\right\}$$

$$\leqslant 2\frac{n+1}{n},$$

with equality, as before, when T is inscribed in the sphere with centre $\mathbf{o}$ and radius 1, and the 'centre of gravity' of T coincides with $\mathbf{o}$. Consequently, we have

$$\frac{\Sigma(T)}{\mu(T)}\geqslant\frac{2\{(n+1)!\}}{\Gamma(\frac{1}{2}n)}\int_0^{\infty}\int_0^{\infty}\dots\int_0^{\infty}\exp\left[-2\frac{n+1}{n}\sum_{1\leqslant\lambda\leqslant\mu\leqslant n}q_{\lambda}q_{\mu}\right]$$

$$\times dq_1dq_2\dots dq_n,$$

with equality in the special case when T is a regular simplex inscribed in the sphere with centre $\mathbf{o}$ and radius 1. Since the right-hand side depends only on n, this proves the lemma.

3. Coverings of space with spheres

It is easy to verify that the $n+1$ spheres of radius 1, with their centres at the vertices of a regular simplex of side

$$\sqrt{\left(\frac{2(n+1)}{n}\right)},$$

just cover the simplex. We define τ_n to be the ratio of the sum of the volumes of the 'sectors' of these spheres, lying in the simplex, to the volume of the simplex.

Now the ratio σ_n, discussed in §7.5, is the ratio of the sum of the volumes of the 'sectors' of the spheres of radius

$$\frac{1}{2}\sqrt{\left(\frac{2(n+1)}{n}\right)},$$

with their centres at the vertices of the simplex, lying in the simplex, to the volume of the simplex. Hence,

$$\tau_n = \left(\frac{2n}{n+1}\right)^{\frac{1}{2}n} \sigma_n. \tag{1}$$

On the other hand, the regular simplex of side

$$\sqrt{\left(\frac{2(n+1)}{n}\right)}$$

is inscribed in the sphere of unit radius, with its centre at the centre of gravity of the simplex. Hence, the minimum value of the ratio $\Sigma(T)/\mu(T)$ of Lemma 1 of §2 above is

$$n\tau_n.$$

We now state and prove the main result of this chapter.

THEOREM 8.1. *If K is an n-dimensional sphere,*

$$\vartheta(K) \geqslant \tau_n. \tag{2}$$

Proof. We suppose that

$$\vartheta(K) < \tau_n$$

and obtain a contradiction. By Theorem 1.9 we may suppose that K is the sphere with radius 1 and centre $\mathbf{o}$. By Theorem 1.9 again, we can find a periodic covering $\mathscr{K}''_P$ of sets

$$K+\mathbf{a}''_i+\mathbf{b}''_j \quad (i = 1, 2, \ldots, N;\ j = 1, 2, \ldots),$$

where $\mathbf{b}''_1, \mathbf{b}''_2, \ldots$, are the points of the lattice Λ'' of points, whose coordinates are integral multiples of s'', such that

$$\rho_-(\mathscr{K}''_P) < \tau_n.$$

Let C'' be a half-open cube, with its edges parallel to the coordinate axes, of edge-length s''. By replacing s'' by one of its integral multiples, if necessary, we can ensure that $s'' > 4$. By Theorem 1.5, we have

$$\frac{N\mu(K)}{\mu(C'')} = \rho_-(\mathscr{K}''_P) < \tau_n.$$

Now take ϵ, with $0 < \epsilon < \frac{1}{2}$, so small that

$$\frac{N\mu(K)}{\mu(\{1-\epsilon\}C'')} < \tau_n.$$

Write

$$C = (1-\epsilon)C'', \quad \Lambda = (1-\epsilon)\Lambda'',$$

$$\mathbf{a}'_i = (1-\epsilon)a''_i \quad (i = 1, 2, \ldots, N),$$

and

$$\mathbf{b}_j = (1-\epsilon)b''_j \quad (j = 1, 2, \ldots).$$

Then $s(C) = (1-\epsilon)s'' > 2$, the system of spheres

$$(1-\epsilon)K+\mathbf{a}'_i+\mathbf{b}_j \quad (i = 1, 2, \ldots, N;\ j = 1, 2, \ldots)$$

forms a periodic covering, and

$$\frac{N\mu(K)}{\mu(C)} < \tau_n.$$

We seek points $\mathbf{a}_1, \mathbf{a}_2 \ldots, \mathbf{a}_N$ satisfying the condition

$$|\mathbf{a}'_i-\mathbf{a}_i| < \epsilon \quad (i = 1, 2, \ldots, N), \tag{3}$$

and certain further conditions. For any choice, subject to (3), the spheres

$$K+\mathbf{a}_i+\mathbf{b}_j \quad (i = 1, 2, \ldots, N;\ j = 1, 2, \ldots)$$

form a periodic covering $\mathscr{K}_P$, with period $s(C)$; to verify this we need merely note that each point of space lies in one of the spheres $(1-\epsilon)K+\mathbf{a}'_i+\mathbf{b}_j$, and so lies in the corresponding sphere $K+\mathbf{a}_i+\mathbf{b}_j$, by virtue of (3). By Theorem 1.5, the density of $\mathscr{K}_P$ is

$$\rho(\mathscr{K}_P) = \frac{N\mu(K)}{\mu(C)} < \tau_n. \tag{4}$$

We investigate the implications of the supposition that there is some point $\mathbf{x}$ of space which is equidistant from $n+2$ points of the system

$$\mathbf{a}_i+\mathbf{b}_j \quad (i = 1, 2, \ldots, N;\ j = 1, 2, \ldots) \tag{5}$$

all lying within unit distance of $\mathbf{x}$. Then, by the periodicity, with period $s(C)$, of the system in each coordinate there is some point $\mathbf{x}$ of C with this property. Suppose then that we have

$$|\mathbf{x}-\mathbf{c}_0| = |\mathbf{x}-\mathbf{c}_1| = \ldots = |\mathbf{x}-\mathbf{c}_{n+1}| < 1, \tag{6}$$

where $\mathbf{c}_0$, $\mathbf{c}_1$, ..., $\mathbf{c}_{n+1}$ are distinct points of the system, say

$$\mathbf{c}_k = \mathbf{a}_{i(k)}+\mathbf{b}_{j(k)} \quad (k = 0, 1, \ldots, n+1).$$

As the points $\mathbf{c}_k$ all lie within the cube concentric with C, of side $2+s(C)$, there are only a finite number of possible choices for the integers $i(0), j(0), \ldots, i(n+1), j(n+1)$. If $i(k)$ had the same value for two different values of k, say k_1 and k_2, we would have

$$|\mathbf{b}_{j(k_1)}-\mathbf{b}_{j(k_2)}| \leqslant |\mathbf{x}-\mathbf{a}_{i(k_1)}-\mathbf{b}_{j(k_1)}| + |\mathbf{x}-\mathbf{a}_{i(k_2)}-\mathbf{b}_{j(k_2)}| < 2.$$

Since $s(C) > 2$, this would imply that $j(k)$ also has the same value, for k equal to k_1 and k_2, so that the corresponding points $\mathbf{c}_k$ coincide. Hence $i(0), i(1), \ldots, i(n+1)$ are necessarily distinct.

Now the equation

$$|\mathbf{x}-\mathbf{c}_k| = |\mathbf{x}-\mathbf{c}_0|$$

takes the scalar product form

$$(\mathbf{x}-\mathbf{c}_0).(\mathbf{c}_k-\mathbf{c}_0) = \tfrac{1}{2}|\mathbf{c}_k-\mathbf{c}_0|^2.$$

Thus the condition (6) implies that

$$\begin{vmatrix} c_1^{(1)}-c_1^{(0)} & \ldots & c_n^{(1)}-c_n^{(0)} & \tfrac{1}{2}|\mathbf{c}_1-\mathbf{c}_0|^2 \\ c_1^{(2)}-c_1^{(0)} & \ldots & c_n^{(2)}-c_n^{(0)} & \tfrac{1}{2}|\mathbf{c}_2-\mathbf{c}_0|^2 \\ \cdots & \cdots & \cdots & \cdots \\ c_1^{(n+1)}-c_1^{(0)} & \ldots & c_n^{(n+1)}-c_n^{(0)} & \tfrac{1}{2}|\mathbf{c}_{n+1}-\mathbf{c}_0|^2 \end{vmatrix} = 0. \tag{7}$$

Since the coefficient of $a_1^{(i(1))}a_2^{(i(2))}\dots a_n^{(i(n))}(a_1^{(i(n+1))})^2$ in the expansion of the left-hand side is $\frac{1}{2}$, this implies a non-trivial restriction on the points $\mathbf{a}_{i(0)}, \mathbf{a}_{i(1)}, \dots, \mathbf{a}_{i(n+1)}$. (Note that this would not necessarily be the case, if some of the points $\mathbf{a}_{i(0)}, \mathbf{a}_{i(1)}, \dots, \mathbf{a}_{i(n+1)}$ could coincide identically. For example, if we had
$i(0) = i(1) = i(2) = i(3)$,

$$\begin{aligned} \mathbf{b}_{j(0)} &= (0, 0, 0, \dots, 0), \\ \mathbf{b}_{j(1)} &= (s, 0, 0, \dots, 0), \\ \mathbf{b}_{j(2)} &= (0, s, 0, \dots, 0), \\ \mathbf{b}_{j(3)} &= (s, s, 0, \dots, 0), \\ s &= s(C), \end{aligned}$$

the determinantal condition would be automatically satisfied.)

For each set of points $\mathbf{a}_1, \mathbf{a}_2, \dots, \mathbf{a}_N$ we consider the point $\mathbf{A}$ in nN-dimensional space, with coordinates

$$(a_1^{(1)}, a_2^{(1)}, \dots, a_n^{(1)}, \dots, a_1^{(N)}, a_2^{(N)}, \dots, a_n^{(N)}).$$

The condition that there is some point $\mathbf{x}$ of space which is equidistant from $n+2$ points of the system (5), all lying within unit distance of $\mathbf{x}$, restricts the point $\mathbf{A}$ to lie on one of a finite number of $(nN-1)$-dimensional surfaces, with equations of the form (7). Henceforth, we suppose that $\mathbf{A}$ has been chosen so that it lies in the nN-dimensional region defined by the inequalities (3), but lies on none of the surfaces. This ensures that the system of points

$$\mathbf{a}_i + \mathbf{b}_j \quad (i = 1, 2, \dots, N; j = 1, 2, \dots) \tag{8}$$

satisfy the two conditions italicized near the beginning of §1 above, with $R = 1$.

We now suppose that the construction described in §1 is applied to the system of points (8). Since this system of points is periodic, in each coordinate, with period $s(C)$, the resultant set V of vertices of the Voronoi polyhedra is periodic in the same way. Thus V is the set of all points

$$\mathbf{v}_k + \mathbf{b}_j \quad (k = 1, 2, \dots, M; j = 1, 2, \dots), \tag{9}$$

where $\mathbf{v}_1, \mathbf{v}_2, \dots, \mathbf{v}_M$ are the points of V that lie in C. The corresponding Delaunay simplices

$$T(\mathbf{v}_k + \mathbf{b}_j) = T(\mathbf{v}_k) + \mathbf{b}_j \quad (k = 1, 2, \dots, M; j = 1, 2, \dots) \tag{10}$$

fit together to fill the whole of space, the only overlappings being of dimension $n-1$.

Hence, as in §7.4, we have

$$\mu(C) = \sum_{k=1}^{M} \sum_{j=1}^{\infty} \mu[C \cap T(\mathbf{v}_k + \mathbf{b}_j)]$$

$$= \sum_{k=1}^{M} \sum_{j=1}^{\infty} \mu[\{C - \mathbf{b}_j\} \cap T(\mathbf{v}_k)]$$

$$= \sum_{k=1}^{M} \mu[T(\mathbf{v}_k)]. \tag{11}$$

Similarly, for any positive ϵ,

$$N\epsilon^n \mu(K) = \sum_{i=1}^{N} \mu(\epsilon K + \mathbf{a}_i)$$

$$= \sum_{i=1}^{N} \sum_{k=1}^{M} \sum_{j=1}^{\infty} \mu[\{\epsilon K + \mathbf{a}_i\} \cap T(\mathbf{v}_k + \mathbf{b}_j)]$$

$$= \sum_{k=1}^{M} \sum_{i=1}^{N} \sum_{j=1}^{\infty} \mu[\{\epsilon K + \mathbf{a}_i - \mathbf{b}_j\} \cap T(\mathbf{v}_k)]$$

$$= \sum_{k=1}^{M} \sum_{i=1}^{N} \sum_{j=1}^{\infty} \mu[\{\epsilon K + \mathbf{a}_i + \mathbf{b}_j\} \cap T(\mathbf{v}_k)]. \tag{12}$$

Now, for fixed k and i and ϵ, with $0 < \epsilon < 1$,

$$\{\epsilon K + \mathbf{a}_i + \mathbf{b}_j\} \cap T(\mathbf{v}_k)$$

is non-empty for at most a finite number of values of j. Further, for fixed i, j, k,

$$\mu[\{\epsilon K + \mathbf{a}_i + \mathbf{b}_j\} \cap T(\mathbf{v}_k)]$$

vanishes for all sufficiently small values of ϵ, except when $\mathbf{a}_i + \mathbf{b}_j$ is one of the vertices of $T(\mathbf{v}_k)$, in which case it is equal to ϵ^n/n times the solid angle of $T(\mathbf{v}_k)$ at the vertex $\mathbf{a}_i + \mathbf{b}_j$. Thus, taking ϵ to be sufficiently small in (12), we obtain

$$nN\mu(K) = \sum_{k=1}^{M} \Sigma[T(\mathbf{v}_k)], \tag{13}$$

where $\Sigma[T(\mathbf{v}_k)]$ is the sum of the solid angles of the simplex $T(\mathbf{v}_k)$. Combining (13), (4) and (11) we see that

$$\sum_{k=1}^{M} \Sigma[T(\mathbf{v}_k)] = nN\mu(K) < n\tau_n \mu(C) = n\tau_n \sum_{k=1}^{M} \mu[T(\mathbf{v}_k)].$$

Hence, for at least one value of k with $1 \leqslant k \leqslant M$,

$$\Sigma[T(\mathbf{v}_k)] < n\tau_n \mu[T(\mathbf{v}_k)]. \tag{14}$$

But, as the spheres of radius 1 with their centres at the points (8) cover the whole of space, and so, in particular, cover the point $\mathbf{v}_k$, it follows from the construction described in §1 that $T(\mathbf{v}_k)$ is contained in the sphere of radius 1 with centre $\mathbf{v}_k$. Hence the inequality (14) is contrary to the remark at the beginning of this section that the ratio $\Sigma(T)/\mu(T)$ of Lemma 1 has minimum $n\tau_n$. This contradiction establishes our theorem that

$$\vartheta(K) \geqslant \tau_n,$$

when K is an n-dimensional sphere.

We remark that, by (1) and Daniels's asymptotic formula,

$$\tau_n \sim \frac{n}{e\sqrt{e}} \quad as \quad n \to \infty. \tag{15}$$

BIBLIOGRAPHY

BAMBAH, R. P.: 1953, On lattice coverings, *Proc. Nat. Inst. Sci. India*, **19**, 447–59; 1954*a*, On lattice coverings by spheres, *Proc. Nat. Inst. Sci. India*, **20**, 25–52; 1954*b*, Lattice coverings with four-dimensional spheres, *Proc. Camb. Phil. Soc.* **50**, 203–8.

BAMBAH, R. P. and DAVENPORT, H.: 1952, The covering of n-dimensional space by spheres, *J. Lond. Math. Soc.* **27**, 224–9.

BAMBAH, R. P. and ROGERS, C. A.; 1952, Covering the plane with convex sets, *J. Lond. Math. Soc.* **27**, 304–14.

BAMBAH, R. P., ROGERS, C. A. and ZASSENHAUS, H.: ——, On coverings with convex domains, submitted to *Acta arith.*

BAMBAH, R. P. and ROTH, K. F.: 1952, Lattice coverings, *J. Indian Math. Soc.* **16**, 7–12.

BARNES, E. S.: 1956, The covering of space by spheres, *Canad. J. Math.* **8**, 293–304; 1957, The complete enumeration of extreme senary forms, *Phil. Trans.* A, **249**, 461–506; 1958, The construction of perfect and extreme forms I, *Acta arith.* **5**, 57–79; 1959, The construction of perfect and extreme forms II, *Acta arith.* **5**, 205–22.

BARNES, E. S. and WALL, G. E.: 1959, Some extreme forms defined in terms of Abelian groups, *J. Aust. Math. Soc.* **1**, 47–63.

BATEMAN, P. T.: 1962, The Minkowski–Hlawka theorem in the Geometry of Numbers, *Arch. der Math.* **13**, 357–62.

BERNSTEIN, B. A.: 1918, Meeting of the San Francisco Section, *Bull. Amer. Math. Soc.* **24**, 417–20 (418).

BLICHFELDT, H. F.: 1914, A new principle in the Geometry of Numbers, with some applications, *Trans. Amer. Math. Soc.* **15**, 227–35; 1919, Report on the Geometry of Numbers, *Bull. Amer. Math. Soc.* **25**, 449–53 (451); 1925, On the minimum value of positive real quadratic forms in 6 variables, *Bull. Amer. Math. Soc.* **31**, 386; 1926, The minimum value of positive quadratic forms in seven variables, *Bull. Amer. Math. Soc.* **32**, 99; 1929, The minimum value of quadratic forms, and the closest packing of spheres, *Math. Ann.* **101**, 605–8; 1934, The minimum values of positive quadratic forms in six, seven and eight variables, *Math. Z.* **39**, 1–15; 1936, A new upper bound to the minimum value of the sum of linear homogeneous forms, *Mh. Math. Phys.* **43**, 410–14.

BLUNDON, W. J.: 1957, Multiple covering of the plane by circles, *Mathematika*, **4**, 7–16; 1963, Multiple packing of circles in the plane, *J. Lond. Math. Soc.* **38**, 176–82.

BONNESEN, T. and FENCHEL, W.: 1934, *Theorie der Konvexen Körper*, *Ergebnisse der Mathematik*, **3**, 1, Springer, Berlin (reprinted in 1948 by Chelsea, New York).

CASSELS, J. W. S.: 1953, A short proof of the Minkowski–Hlawka theorem, *Proc. Camb. Phil. Soc.* **49**, 165–6; 1959, An introduction to the Geometry of Numbers, *Grundl. Math. Wiss.* **99**, Springer, Berlin.

CHALK, J. H. H.: 1950, On the frustrum of a sphere, *Ann. Math.* (2), **52**, 199–216.

CHALK, J. H. H. and ROGERS, C. A.: 1948, The critical determinant of a convex cylinder, *J. Lond. Math. Soc.* **23**, 178–87.

CHAUNDY, T. W.: 1946, The arithmetic minima of positive quadratic forms, *Quart. J. Math.* **17**, 166–92, but see also the review: Coxeter, H. S. M., 1947, *Math. Rev.* **8**, 137–8.

COXETER, H. S. M.: 1958, Close-packing and froth, *Illinois J. Math.* **2**, 746–58; 1962*a*, The classification of zonohedra by means of projective diagrams, *J. Math. pures appl.*, (9), **41**, 137–56; 1962*b*, The problem of packing a number of equal nonoverlapping circles on a sphere, *Trans. N.Y. Acad. Sci.* II, **24**, 320–31.

COXETER, H. S. M., FEW, L. and ROGERS, C. A.: 1959, Covering space with equal spheres, *Mathematika*, **6**, 147–57.

COXETER, H. S. M. and TODD, J. A.: 1951, An extreme duodenary form, *Canad. J. Math.* **5**, 384–92.

DAVENPORT, H.: 1951, Sur un système de sphères qui recouvrent l'espace à n dimensions, *C.R. Acad. Sci., Paris*, **233**, 571–3; 1952, The covering of space by spheres, *R.C. Circ. mat. Palermo*, (2), **1**, 92–107; 1955, Problèmes d'empilement et de découvrement, Teoria dei numeri, *Centro Internazionale Matematico Estivo, 2° Ciclo, Varenna, Villa Monastero*, 16–25 agosto, 1955.

DAVENPORT, H. and ROGERS, C. A.: 1947, Hlawka's theorem in the Geometry of Numbers, *Duke Math. J.* **14**, 367–75.

DAVIES, H. L.: —— Thesis in course of preparation for submission to the University of London.

DELAUNAY, B.: 1934, Sur la sphère vide, *Bull. Acad. Sci. U.S.S.R.* (VII), *Classe Sci. Mat. Nat.* 793–800.

ENNOLA, V.: 1961, On the lattice constant of a symmetric convex domain, *J. Lond. Math. Soc.* **36**, 135–8.

ERDŐS, P. and ROGERS, C. A.: 1953, The covering of n-dimensional space by spheres, *J. Lond. Math. Soc.* **28**, 287–93; 1962, Covering space with convex bodies, *Acta arith.* **7**, 281–5.

ESTERMANN, T.: 1928, Über den Vektorenbereich eines konvexen Körpers, *Math. Z.* **28**, 471–5.

FÁRY, I.: 1950, Sur la densité des réseaux de domaines convexes, *Bull. Soc. Math. Fr.* **78**, 152–61.

FEJES TÓTH, J.†: 1940, Über einen geometrischen Satz, *Math. Z.* **46**, 79–83; 1943, Über die dichteste Kugellagerung, *Math. Z.* **48**, 676–84; 1946, Eine Bemerkung über die Bedeckung der Ebene durch Eibereiche mit Mittelpunkt, *Acta Sci. Math. Szeged*, **11**, 93–5; 1948, On the densest packing of convex domains, *Proc. K. Akad. Wet. Amsterdam*, **51**, 544–7; 1950, Some packing and covering theorems, *Acta Sci. Math. Szeged*, 12/A, 62–7; 1952, Über das Problem der dichtesten Kugellagerung, *C.R. d. 1° Congrès d. Math. Hongrois*, 1950, Akad. Kiadó, Budapest, pp. 619–42; 1953*a*,

† The earlier papers appeared under the name L. Fejes.

Lagerungen in der Ebene, auf der Kugel und in Raum, *Grundl. Math. Wiss.* **65**, Springer, Berlin; 1953*b*, On close-packing of spheres in spaces of constant curvature, *Publ. Math. Debrecen*, **3**, 158–67; 1956*a*, On the volume of a polyhedron in non-Euclidean spaces, *Publ. Math. Debrecen*, **4**, 256–61; 1956*b*, Characterisation of the nine regular polyhedra by extremum properties, *Acta Math. Acad. Sci. Hungary*, **7**, 31–48; 1957, Filling of a domain by isoperimetric discs, *Publ. Math. Debrecen*, **5**, 119–27; 1959*a*, Über eine Punktverteilung auf der Kugel, *Acta Math. Acad. Sci. Hungary*, **10**, 13–19; 1959*b*, Annäherung von Eibereichen durch Polygone, *Math.-Phys. Semesterberichte*, **6**, 253–61; 1959*c*, An extremal distribution of great circles on a sphere, *Publ. Math. Debrecen*, **6**, 79–82; 1959*d*, Kugelunterdeckungen und Kugelüberdeckungen in Räumen konstanter Krümmung, *Arch. Math.* **10**, 307–313.

Fejes Tóth, L. and Heppes, A.: 1960, Filling of a domain by equiareal discs, *Publ. Math. Debrecen*, **7**, 198–203.

Fejes Tóth, L. and Molnár, J.: 1958, Unterdeckung und Überdeckung der Ebene durch Kreise, *Math. Nachr.* **18**, 235–43.

Few, L.: 1953, The double packing of spheres, *J. Lond. Math. Soc.* **28**, 297–304; 1956, Covering space by spheres, *Mathematika*, **3**, 136–9; 1960, A mixed packing problem, *Mathematika*, **7**, 56–63; — , Multiple packing of spheres, to appear in *J. Lond. Math. Soc.*

Florian, A.: 1960, Ausfüllung der Ebene durch Kreise, *R.C. Circ. mat. Palermo*, (2), **9**, 1–13; 1961, Überdeckung der Ebene durch Kreise, *R.C. Sem. mat. Padova*, **31**, 77–86; 1962, Zum Problem der dunnsten Kreisüberdeckung der Ebene, *Acta Math. Acad. Sci. Hungary*, **13**, 397–400.

Gauss, C.F.: 1831, Untersuchungen über die Eigenschaften der positiven ternären quadratischen Formen von Ludwig August Seeber, *Göttingische gelehrte Anzeigen*, 1831 Juli 9, see *Werke*, Göttingen, 1836, ii, 188–96, or *J. reine angew. Math.* **20**, 1840, 312–20.

Grünbaum, B.: ——, On a characterization of simplices, privately circulated note.

Hancock, H.: 1939, *Development of the Minkowski Geometry of Numbers*, Macmillan, New York.

Heppes, A.: 1955, Über mehrfache Kreislagerungen, *Elem. Math.* **10**, 125–7; 1959, Mehrfache gitterförmige Kreislagerungen in der Ebene, *Acta Math. Acad. Sci. Hungary*, **10**, 141–8.

Heppes, A. and Molnár, J.: 1960, Újabb eredmények a diszkrét geometriában, *Mat. Lapok*, **11**, 330–55; 1962, Újabb eredmények a diszkrét geometriaban II, *Mat. Lapok*, **13**, 39–72.

Hlawka, E.: 1944, Zur Geometrie der Zahlen, *Math. Z.* **49**, 285–312; 1945, Über Potenzsummen von Linearformen, *S.B. Akad. Wiss. Wien, Math.-nat. Kl.* (II*a*), **154**, 50–8; 1947, Über Potenzsummen von Linearformen II, *S.B. Akad. Wiss. Wien, Math.-nat. Kl.* (II*a*), 247–54; 1948, Ausfüllung und Überdeckung durch Zylinder, *Anz. öst. Akad. Wiss. Wien, Math.-Nat. Kl.* **85**,

116–19; 1949, Ausfüllung und Überdeckung konvexer Körper durch konvexe Körper, *Mh. Math. Phys.* **53**, 81–131.

KELLER, O.-H.: 1954, Geometrie der Zahlen, *Enzykl. math. Wiss.*, Band 1_2, Heft 11, Teil 3, B. G. Teubner, Leipzig.

KERSHNER, R.: 1939, The number of circles covering a set, *Amer. J. Math.* **61**, 665–71.

KORKINE, A. and ZOLOTAREFF, G.: 1872, Sur les formes quadratique positive quaternaires, *Math. Ann.* **5**, 581–3; 1877, Sur les formes quadratiques positives, *Math. Ann.* **11**, 242–92.

LAGRANGE, J. L.: 1773, Recherches d'arithmétique, *Nouveaux Mémoires de l'Académie royal des Sciences et Belles-Lettres de Berlin*, pp. 265–312, Oeuvres, III, 693–758.

LEKKERKERKER, C. G.: 1956, On the Minkowski–Hlawka theorem, *Proc. K. Ned. Akad. Wet.* A, **59**, 426–34.

MACBEATH, A. M.: 1951, A compactness theorem for convex regions, *Canad. J. Math.* **3**, 54–61.

MACBEATH, A. M. and ROGERS, C. A.: 1955, A modified form of Siegel's mean value theorem, *Proc. Camb. Phil. Soc.* **51**, 565–76; 1958*a*, Siegel's mean value theorem in the geometry of numbers, *Proc. Camb. Phil. Soc.* **54**, 139–51; 1958*b*, A modified form of Siegel's mean value theorem, *Proc. Camb. Phil Soc.* **54**, 322–5.

MAHLER, K.: 1944*a*, On a theorem of Minkowski, *J. Lond. Math. Soc.* **19**, 201–5; 1946*a*, On lattice points in a cylinder, *Quart. J. Math.* **17**, 16–18; 1946*b*, The theorem of Minkowski–Hlawka, *Duke Math. J.* **13**, 611–21; 1947*a*, On the area and the densest packing of convex domains, *Proc. K. Ned. Akad. Wet. Amsterdam*, **50**, 109–18; 1947*b*, On the minimum determinant and the circumscribed hexagons of a convex domain, *Proc. K. Ned. Akad. Wet. Amsterdam*, **50**, 692–703.

MALYŠEV, A. V.: 1952, On the Minkowski–Hlawka theorem concerning a star body, *Usp. Matem. Nauk* (N.S.) **7**, no. 2 (48), 168–71 (in Russian).

MINKOWSKI, H.: 1893*a*, Extrait d'une lettre adressée à M. Hermite, *Bull. Sci. math.* (2), **17**, 24–9; 1893*b*, Über Eigenschaften von ganzen Zahlen, die durch räumliche Anschauung erschlossen sind, *Mathematical papers read at the International Mathematical Congress held in connection with the World's Columbian Exposition, Chicago*, 1893, pp. 201–7, see H. Minkowski (1911), I, pp. 271–7; 1904, Dichteste gitterförmige Lagerung kongruenter Körper, *Nachr. Ges. Wiss. Göttingen*, 311–55; 1905, Diskontinuitätsbereich für arithmetische Aequivalenz, *J. reine angew. Math.* **129**, 220–74; 1911, *Gesammelte Abhandlungen*, Teubner, Berlin.

MOLNÁR, J.: 1961, Alcune generalizzazioni del teorema di Segre–Mahler, *Accad. Naz. Lincei Rend., Cl. Sci. fis. mat. nat.* (8), **30**, 700–5.

MORDELL, L. J.: 1944, Observation on the minimum of a positive quadratic form in eight variables, *J. Lond. Math. Soc.* **19**, 3–6.

OLER, N.: 1961, An inequality in the geometry of numbers, *Acta math.* **105**, 19–48.

Rademacher, H.: 1925, Über den Vektorenbereich eines konvexen ebenen Bereichs, *Jber. dtsch. Mat. Ver.* **34**, 64–79.

Rado, R.: 1949, Some covering theorems (I), *Proc. Lond. Math. Soc.* (2), **51**, 232–64; 1951, Some covering theorems (II), *Proc. Lond. Math. Soc.* (2), **53**, 243–67.

Rankin, R. A.: 1947, On the closest packing of spheres in *n* dimensions, *Ann. Math.* (2), **48**, 1062–81; 1948, On sums of powers of linear forms III, *Proc. K. Ned. Akad. Wet. Amsterdam*, **51**, 846–53; 1949*a*, On sums of powers of linear forms I, *Ann. Math.* (2), **50**, 691–8; 1949*b*, On sums of powers of linear forms II, *Ann. Math.* (2), **50**, 699–704; 1955, The closest packing of spherical caps in *n* dimensions, *Proc. Glasgow Math. Ass.* **2**, 139–44.

Reinhardt, K.: 1934, Über die dichteste gitterförmige Lagerung kongruenter Bereiche in der Ebene und eine besondere Art konvexer Kurven, *Abh. math. Sem. hansische Univ.* **10**, 216–230.

Rogers, C. A.: 1947, Existence theorems in the Geometry of Numbers, *Ann. Math.* (2), **48**, 994–1002; 1950, A note on coverings and packings, *J. Lond. Math. Soc.* **25**, 327–31; 1951*a*, The number of lattice points in a star body, *J. Lond. Math. Soc.* **26**, 307–10; 1951*b*, The closest packing of convex two-dimensional domains, *Acta math.* **86**, 309–21; 1953*a*, Certain integrals over convex sets, *J. Lond. Math. Soc.* **28**, 293–7; 1953*b*, The volume of a polyhedron inscribed in a sphere, *J. Lond. Math. Soc.* **28**, 410–16; 1954, The Minkowski–Hlawka theorem, *Mathematika*, **1**, 111–24; 1955*a*, The moments of the number of points of a lattice in a bounded set, *Phil. Trans.* A, **248**, 225–51; 1955*b*, Mean values over the space of lattices, *Acta Math.* **94**, 249–87; 1956*a*, The number of lattice points in a set, *Proc. Lond. Math. Soc.* (3), **6**, 305–20; 1956*b*, Two integral inequalities, *J. Lond. Math. Soc.* **31**, 235–8; 1957*a*, A note on coverings, *Mathematika*, **4**, 1–6; 1957*b*, A single integral inequality, *J. Lond. Math. Soc.* **32**, 102–8; 1958*a*, Lattice coverings of space with convex bodies, *J. Lond. Math. Soc.* **33**, 208–12; 1958*b*, Lattice coverings of space: the Minkowski–Hlawka theorem, *Proc. Lond. Math. Soc.* (3), **8**, 447–65; 1958*c*, The packing of equal spheres, *Proc. Lond. Math. Soc.* (3), **8**, 609–20; 1959, Lattice coverings of space, *Mathematika*, **6**, 33–9; 1960, The closest packing of convex two-dimensional domains, corrigendum, *Acta Math.* **104**, 305–6.

Rogers, C. A. and Shephard, G. C.: 1957, The difference body of a convex body, *Arch. Math.* **8**, 220–33; 1958, Convex bodies associated with a given convex body, *J. Lond. Math. Soc.* **33**, 270–81.

Sanov, I. N.: 1952, A new proof of Minkowski's theorem, *Izv. Akad. Nauk, SSSR Ser. Mat.* **16**, 101–112 (in Russian).

Santaló, L. A.: 1950, Integral geometry in projective and affine spaces, *Ann. Math.* (2), **51**, 739–55.

Sas, E.: 1939, Über eine Extremumeigenschaft der Ellipsen, *Comp. Math.* **6**, 468–70.

Schmidt, W.: 1956*a*, Eine neue Abschätzung der kritischen Determinante von Sternkörpern, *Mh. Math.* **60**, 1–10; 1956*b*, Eine Verschärfung des Satzes von Minkowski–Hlawka, *Mh. Math.* **60**, 110–13; 1957, Mittelwerte über Gitter, *Mh. Math.* **61**, 269–76; 1958*a*, Mittelwerte über Gitter, II, *Mh. Math.* **62**, 250–8; 1958*b*, The measure of the set of admissible lattices, *Proc. Amer. Math. Soc.* **9**, 390–403; 1959, Masstheorie in der Geometrie der Zahlen, *Acta Math.* **102**, 159–224; 1963, On the Minkowski–Hlawka Theorem, *Illinois J. Math.* **7**, 18–23.

SEEBER, L. A.: 1831, *Untersuchungen über die Eigenschaften der positiven ternären quadratischen Formen*, Freiburg, 248 pp.

SEGRE, B. and MAHLER, K.: 1944, On the densest packing of circles, *Amer. Math. Mon.* **51**, 261–70.

SIEGEL, C. L.: 1945, A mean value theorem in the geometry of numbers, *Ann. Math.* (2), **46**, 340–7.

SMITH, N. E.: 1951, On a packing problem of statistical number geometry, Ph.D. thesis, McGill University.

SÜSS, W.: 1928, Über den Vektorenbereich eines Eikörpers, *Jber. dtsch. Mat.Ver.* **37**, 359–62.

THUE, A.: 1882, Om nogle geometrisk-taltheoretiske Theoremer, *Forhandlingerne ved de Skankinaviske Naturforskeres*, 14 *Møde, Kjøbenhavn*, 352–53, in Danish; 1910, Über die dichteste Zusammenstellung von kongruenten Kreisen in einer Ebene, *Skr. Vidensk-Selsk., Christ.*, No. 1, pp. 1–9.

VORONOI, G. (spelt Voronoï in the French original): 1908, Nouvelles applications des paramètres continus à la théorie des formes quadratiques, Deuxième Mémoire, Recherches sur les parallélloèdres primitifs, *J. reine angew. Math.* **134**, 198–287.

WATSON, G. L.: 1956, The covering of space by spheres, *R.C. Circ. mat. Palermo*, (2), **5**, 93–100.

WEIL, A.: 1946, Sur quelques résultats de Siegel, *Summa bras. Math.* **1**, 21–34.

WHITWORTH, J. V.: 1948, On the densest packing of sections of a cube, *Ann. Mat. pura appl.* (4), **27**, 29–37; 1951, The critical lattices of the double cone, *Proc. Lond. Math. Soc.* (2), **53**, 422–43.

WOODS, A. C.: 1958, The critical determinant of a spherical cylinder, *J. Lond. Math. Soc.* **33**, 357–68.

YEH, Y.: 1948, Lattice points in a cylinder over a convex domain, *J. Lond. Math. Soc.* **23**, 188–95.

ZASSENHAUS, H.: 1961, Modern developments in the geometry of numbers, *Bull. Amer. Math. Soc.* **67**, 427–39.

INDEX